It's About Time

LIZ EVERS

It's About Time

From CALENDARS and CLOCKS to
MOON CYCLES and LIGHT YEARS
– A HISTORY

Michael O'Mara Book Limited

First published in Great Britain in 2013 by
Michael O'Mara Books Limited
9 Lion Yard
Tremadoc Road
London SW4 7NQ

A CIP catalogue record for this book is available from the
British Library.

Papers used by Michael O'Mara Books Limited are natural,
recyclable products made from wood grown in sustainable
forests. The manufacturing processes conform to the
environmental regulations of the country of origin.

ISBN: 978-1-78243-067-4 in hardback print format
ISBN: 978-1-78243-193-0 in paperback print format
ISBN: 978-1-78243-087-2 in EPub format
ISBN: 978-1-78243-088-9 in Mobipocket format

1 2 3 4 5 6 7 8 9 10

Cover design by Ana Bjezancevic
Designed and typeset by K DESIGN, Somerset
Illustrations by Greg Stevenson
Maps on pages 108 and 125 by David Woodroffe

Printed and bound by
CPI Group (UK) Ltd, Croydon, CR0 4YY

www.mombooks.com

Contents

Acknowledgements

My sincerest thanks to all at Michael O'Mara Books, especially to my editor Anna Marx for her support, enthusiasm and input, Ana Bježančević for her ever-lovely design work, and Greg Stevenson for his wonderful illustrations. Thanks also to Dan O'Grady and my brother Peter Evers for all their useful suggestions. And finally the biggest thank you goes to my great friend and editor-extraordinaire, Silvia Crompton, for all the precious time she has given to me these past few years.

Introduction

A few years ago some startling images were captured by Brazil's Indian Affairs Department. Taken from a plane flying high above the Amazon near the border of Brazil and Peru, the images showed members of an 'uncontacted' tribe. Some were painted red, others black, but all were looking up curiously at the metal bird cutting through the sky above.

Looking at these images felt a bit like time travel; looking at the past in the present, or two dimensions co-existing. These people do not know that it's the 'twenty-first century'. To them, we are the weird creatures from another time, possibly even another world. How long this 'past' in the Amazon can

9

continue is uncertain, as modern man encroaches ever more into these ancient tribal lives, sometimes violently, in the name of progress.

A few months after these images entered the public domain I came across another story, this time about a recently contacted tribe, the Amondawa in Brazil. First 'discovered' by anthropologists in 1986, the Amondawa do not have an abstract concept of time. They have no word for time, or divisions of time – such as months or years. Rather than talk about age they assign different names to each other to indicate the different stages of their lives or their status within their community. They have no 'time technology' – no calendars or clocks – and only a limited numbering system.

What struck me is just how difficult this kind of life is to comprehend. And I realized how obsessed we modern people are with time – especially not having enough of it – and just how unique the Amondawa are in the absence of this obsession. I also realized how little I understood about time, how we capture and create it, and how our Earth and our bodies interact with it.

We each live in our own psychological time – memories of the past, anticipation for the future – and these 'time zones' co-exist with our present, our now. And we experience time subjectively – an hour is a long time in a doctor's waiting room, but can fly by with good friends.

This book goes back to the beginning of time as we know it – right back to the beginning of the universe and starts from there. It pieces together the history of time as perceived and processed by our forebears and by the great scientific minds of our current age – and it also tries to have a little fun along the way.

We'll journey through geological ages, meet dinosaurs and distant cousins, tell time by the Moon and the Sun, and learn about the clocks within us which dictate the rhythms of our daily lives. We'll look at the evolution of time technologies, from the earliest calendars etched on the bones of eagles' wings to quantum clocks. We'll see how time is speeding up and how it's slowing down, we'll travel into the future through wormholes and black holes, span light years and peek into parallel dimensions. And for aspiring time travellers, there will be tips and tricks for journeying into the past and future along the way.

1
The Land Before Time

Happy Birthday Planet Earth

In 1654, the Anglican Bishop of Armagh, James Usher, announced that the universe was created at six o'clock on the evening of 22 October 4004 BCE. He reportedly came to this rather definitive conclusion after years of studying the Bible and world history. This theory of the Earth's age was pretty popular right up to the nineteenth century, when the study of geology and Darwin's theory of evolution made it clear that the world was considerably older.

It is now widely believed to be 4.54 billion years old – or written out in full – 4,540,000,000 years old. That's a lot of years. The 4.54 billion figure has

13

been reached using rather complex mathematics combined with the methods of 'radiometric' dating – which include radiocarbon dating, potassium-argon dating and uranium lead dating.

At its most basic, radiometric dating looks at radioactive decay. It compares the amount of a naturally occurring radioactive chemical component (isotope) and its decay products – we know, for example, that the radioactive component uranium decays to become lead, so looking at the amount of lead left in a rock one can calculate how much uranium there would have been to start with and so how long it has taken to produce the lead.

Applying these techniques to really, really old rocks and minerals – including meteorites and lunar samples – the magic figure of 4.54 billion has been reached and agreed upon. For now.

The oldest known terrestrial materials are zircon crystals found in Western Australia. These have been dated as over 4.4 billion years old. The oldest known meteorite matter is 4.567 billion years old. It is believed that our solar system can't be much older than these samples.

Which brings us to the time before there was an Earth, or a solar system to house it. To when our

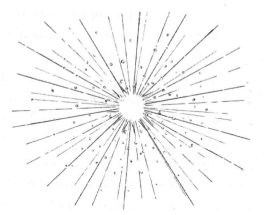

*The Big Bang is dated as starting
13.5 and 13.75 billion years ago*

universe was born. The prevailing theory is that of the Big Bang, when the universe started expanding from a dense and hot state – and continues to expand into space, which is itself continually expanding.

The geologic time scale

Coming back down to Earth again, something called the 'geologic' time scale is used by earth scientists, geologists and palaeontologists to describe timings and events in our Earth's past. It relates time to 'stratigraphy' – the study of layers of rocks (stratification).

There are many wonderful examples of stratification bearing testament to the Earth's long history. Examples are found in chalk layers in Cyprus, the stunning Colorado Plateau in Utah, exposed strata on mountain faces in the French Alps, and the amazing Stratified Island near La Paz, Mexico, to name but a few.

The units used to describe geologic time are very long. They include Eons (half a billion years), Eras (several hundred million years), Epochs (tens of millions of years), and Ages (millions of years).

Taking it as read that the Earth is 4.54 billion years old, the deposits of our old pal zircon, the oldest known mineral, were found during the Hadean Eon in the Cryptic Era. This is when the Moon and Earth were formed. Between 500 and 600 million years later in the Eoarchean Era, simple single-celled life came into being, evidence for which is found in microfossils – that is, fossils which are not larger than four millimetres, and often smaller than one millimetre, and which can only be studied using light or electron microscopy.

Skipping ahead to the Proterozoic Eon, geologic evidence shows that our atmosphere became oxygenic (specifically during the Palaeoproterozoic

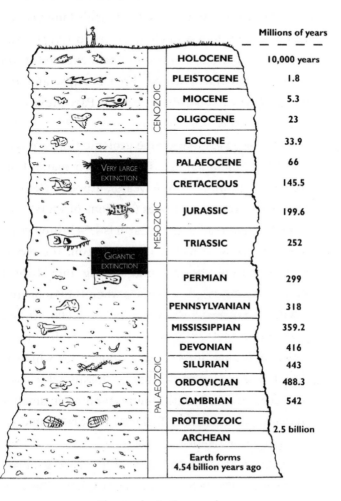

			Millions of years
		HOLOCENE	10,000 years
	CENOZOIC	PLEISTOCENE	1.8
		MIOCENE	5.3
		OLIGOCENE	23
		EOCENE	33.9
VERY LARGE EXTINCTION		PALAEOCENE	66
	MESOZOIC	CRETACEOUS	145.5
		JURASSIC	199.6
GIGANTIC EXTINCTION		TRIASSIC	252
	PALAEOZOIC	PERMIAN	299
		PENNSYLVANIAN	318
		MISSISSIPPIAN	359.2
		DEVONIAN	416
		SILURIAN	443
		ORDOVICIAN	488.3
		CAMBRIAN	542
		PROTEROZOIC	2.5 billion
		ARCHEAN	
		Earth forms 4.54 billion years ago	

The geologic time scale

17

Era some 2.05 billion years ago), then the first complex single-celled life, protists, came into being around 1.8 billion years ago.

It took another 1.2 billion years for the first fossils of multi-celled animals (worms, sponges, soft jelly-like creatures) to show up during the Neo-proterozoic Era (around 635 million years ago) and these evolved into yet more complex fishy creatures during the long Palaeozoic Era (between 541 and 255 million years ago). By the end of this Era the landmass known as Pangaea had formed, comprised of North America, Europe, Asia, South America, Africa, Antarctica, and Australia. Various reptiles and amphibians were roaming about and basic flora, mosses and primitive seed plants had developed, while a host of marine life flourished in shallow reefs.

Thence to the Mesozoic Era. During its Triassic, Jurassic and Cretaceous Periods (between 252 and 72 million years ago) the dinosaurs, first mammals and crocodilia appeared. Then flowering plants and all manner of new types of insects. Towards the end of the Cretaceous Period there were many new species of dinosaur (though not for long) and creatures equivalent to modern crocodiles and sharks. Primitive birds replaced pterosaurs and the first marsupials appeared. Plus atmospheric CO_2 was close to our present-day levels.

Which brings us to our own Era – the Cenozoic – which started some 66 million years ago and is often referred to as the 'age of mammals'. In the early part of this Era, the dinosaurs were extinct (more on this to follow) and mammals were diversifying, but it would still be another 40-plus million years before the first apes, our evolutionary ancestors, appeared.

And it wasn't until just 200,000 years ago that the first anatomically modern humans appeared and only 50,000 years ago during the Holocene Epoch (which we're still in) that we started tinkering with stone tools.

The bottom line is the Earth is very old, and we are very young upon it. To put things in perspective, if you think of the age of the Earth as a 24-hour clock, the first humans appear just 40 seconds before midnight at 23:59:20.

So what happened to the dinosaurs?

It is now generally agreed that the catchily titled 'Cretaceous-Palaeogene extinction event', which happened approximately 65.5 million years ago, led to the mass extinction of the dinosaurs.

However, the actual nature of the event is still a matter of considerable discussion. Theories range from a massive asteroid or meteor impact to increased volcanic activity altering the biosphere and significantly reducing the amount of sunlight reaching Earth.

Whatever it was, the event left behind a geological signature known variously as the Cretaceous-Palaeogene boundary, K-T boundary or K-Pg boundary. Non-avian dinosaurs were wiped out, their fossils lying below the boundary, indicating they became extinct during the event. The small number of dinosaur fossils that have been found above the boundary have been explained as having eroded from their original positions and preserved in later sedimentary layers. You can see exposed areas of the boundary in wilderness areas and state parks, such as at Trinidad Lake, Colorado, and Drumheller in Alberta, Canada.

Ice ages

Technically, we are still in an ice age. Admittedly at the tail end of it, the worst was over around 12,500 years ago. It began 2.6 million years ago, but the presence of ice sheets in Greenland and the Antarctic signal its continued existence.

The Swiss geographer and engineer Pierre Martel (1706–1767) was the first to posit the theory of ice ages. On a visit to the Chamonix valley in the Alps, he observed that the dispersal of boulders pointed to the fact that the glaciers had once been much larger, but had contracted with time. And this phenomenon was observable in other parts of Switzerland, Scandinavia and later noted in the Chilean Andes. But it wasn't until the 1870s that the theory was widely accepted as fact.

In addition to the erratic dispersal of large boulders, other evidence of ice ages comes in the form of rock scouring and scratching, valley cutting, the creation of small hills called drumlins and unusual patterns in the distribution of fossils.

There have been at least five ice ages in our Earth's history – and outside of these ages the Earth appears

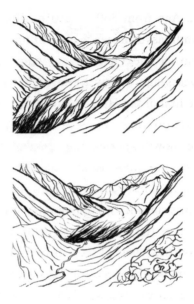

Contraction of a glacier in Chamonix

to have been free of ice, even at high latitudes. The first ice age was the Huronian, which is thought to have extended from 2.4 billion years ago to 2.1 billion years ago (that's before the existence of complex single-celled life forms). This was followed by the Cryogenian from 850 to 635 million years ago (when multi-celled creatures were evolving); the relatively short Andean-Saharan from 460 to 430 million years ago (as more complex marine life was evolving); the

Karoo Ice Age from 360 to 260 million years ago (as the landmass Pangaea was forming); and finally the current ice age, Quaternary, which started 2.58 million years ago (a few hundred thousand years before the first of the Homo genus had evolved) and continues to this day.

We are now experiencing a relatively stable 'interglacial' period, which has provided the climate conditions that have allowed our race to flourish. Without this stability we may not have survived.

As to when the next ice age begins in earnest depends on the levels of CO_2 in the atmosphere. A sudden drop would speed up the arrival of the next ice age – even as soon as 15,000 years hence. But estimates based on rising CO_2 (the more likely case given our penchant for fossil fuels) suggest that our current interglacial period may persist for another 50,000 years or even considerably longer.

Human evolution

It is astonishing how recent is most of our knowledge about ourselves and our planet. As mentioned above, the concept of ice ages was only first posited in the mid-1700s and generally accepted in the 1870s.

The ideas of the 'evolution' of species, including humans, and 'natural selection' have only been knocking around since the mid-1800s, and only brought to the fore in 1859 when Charles Darwin (1809–1882) published *On the Origin of Species*. Even so, it took many more decades for Darwin's ideas about evolution to become mainstream and be incorporated into life sciences. Thinking about the Earth's age as a 24-hour clock again, the most infinitesimal units of time measurement would be required to place these discoveries in our planet's natural history.

There was uproar just a century and half ago when Darwin more explicitly outlined his theories about human evolution in his seminal 1871 book *The Descent of Man, and Selection in Relation to Sex*. In it he suggests that human races evolved from a common ancestor – and that common ancestor from a succession of animals over millennia. It was an idea appalling to the majority of the day.

But with close study and uncovering ever more substantive evidence, the proof for many of Darwin's ideas became too compelling to deny. The evolutionary theory that emerged is now widely accepted as fact by the scientific community, if not by various religious communities. To this day 'Creationists', like our friend James Usher (1581-1656), Bishop of Armagh, believe the world was created by God in six days around 4004 BCE.

Our now considerable knowledge of the geologic time scale and fossil records gives us a fascinating portrait of the development of life on Earth. And discoveries in archaeology, palaeontology and DNA research continue to provide a vivid picture of our evolution as a species. We've seen already that it was just 200,000 years ago that the first anatomically modern humans appeared.

The Homo genus

It is thought that primates, from whom humans are descended, diverged from other mammals about 85 million years ago, though the earliest fossil records we have are from around 55 million years ago. The first bipeds diverged around 4 to 6 million years ago, splitting from cousin primates like chimpanzees with whom we share a common ancestor, and eventually evolving into the genus, or biological classification, Homo. There is no definitive timeline for the Homo genus, and there are many candidates for the evolutionary links in our chain:

HOMO HABILIS
(3 TO 2 MILLION YEARS AGO)

The first documented members of the genus Homo, *Homo habilis* evolved around 2.3 million years ago in South and East Africa. It is thought to be the earliest species to use stone tools. *Homo habilis*'s brains were around the size of a chimpanzee's. In May 2010, a new species, *Homo gautengensis,* was discovered in South Africa and may have evolved earlier than *Homo habilis*, but this has yet to be agreed conclusively.

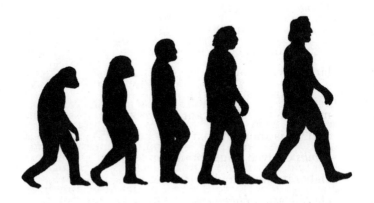

HOMO RUDOLFENSIS AND HOMO GEORGICUS
(1.9 TO 1.6 MILLION YEARS AGO)

These are proposed species names for fossils from about 1.9 to 1.6 million years ago but whose relation to *Homo habilis* is not yet clear. There is just one *Homo rudolfensis* specimen – an incomplete skull from Kenya, which may or may not be another *Homo habilis*. *Homo georgicus* comes from Georgia, in the Caucasus region, and may be an intermediate form between *Homo habilis* and *Homo erectus*.

HOMO ERECTUS
(1.8 MILLION TO 70,000 YEARS AGO)

Homo erectus had a long evolutionary lifespan. Records indicate that the species lived from about 1.8 million to about 70,000 years ago, possibly being largely wiped out by the so-called Toba catastrophe (a volcanic super-eruption in Indonesia, where many of the significant *Homo erectus* fossil finds are). It is thought that some populations of *Homo habilis* evolved larger brains and started to use more elaborate stone tools – leading to the new advanced classification *Homo erectus*. Other key physiological changes include the evolution of locking knees and a different location of the foramen magnum (the hole in the skull where the spine enters).

HOMO HEIDELBERGENSIS
(800,000 TO 300,000 YEARS AGO)

Homo heidelbergensis (also 'Heidelberg Man', after the University of Heidelberg) could be the direct ancestor of both *Homo neanderthalensis* (Neanderthals, see page 29) in Europe and *Homo sapiens*. The missing link, if you will. The best evidence found for these hominines dates them to between 600,000 and 400,000 years ago, but it is thought that they may have lived from about 800,000 to about 300,000 years ago.

Homo heidelbergensis used stone-tool technology that was very close to those used by *Homo erectus*, and recent findings of twenty-eight skeletons in Atapuerca in Spain suggest that this species may have been the first of the Homo genus to bury their dead. It is also thought that *Homo heidelbergensis* may have had a primitive form of language, although no forms of art (often equated with symbolic thinking and language) have been uncovered in relation to this species.

HOMO SAPIENS
(250,000 TO 200,000 YEARS AGO TO PRESENT)

The most important evolutionary period for our species occurred between 400,000 and 250,000 years ago – the period of transition from *Homo erectus* to *Homo sapiens*. During this time, our cranial sizes expanded, meaning bigger brains, and we began to use ever-increasingly elaborate stone tools. As a species *Homo sapiens* are highly homogenous, genetically speaking. This is relatively unusual in any species so widely disbursed and is seen as evidence that we evolved in a particular place (Africa) and migrated from there. But we have evolved certain region-specific adaptive traits such as skin colour, and eyelid and nose shapes, for example.

Neanderthal man

Named after the Neander Valley in Germany where the species was first discovered, Neanderthals are alternatively classified as a subspecies of *Homo sapiens* or as a separate species but of the same Homo genus.

The earliest Neanderthals are though to have appeared in Europe 600,000 to 350,000 years ago (no evidence of Neanderthals has been found in Africa) – and to have survived there until around 25,000 years ago. Often characterized as primitive creatures with low brows and weak chins, they in fact used advanced tools (projectile points, bone tools), had a language and lived in complex social groups. The Neanderthal cranial capacity is thought to have been the same size as modern humans, possibly bigger. And when it comes to brains in the Homo genus, size really does matter.

Neanderthals disappeared from the fossil record about 25,000 years ago. Theories abound as to what happened to them. But apart from hypotheses about a volcanic 'super-eruption' or their slowness to adapt to rapid changes in climate leading to their demise, it seems that the worst thing for Neanderthals was us. It is thought that Neanderthals were most likely driven to extinction because of living in competition with ever-expanding human populations. However, there is also evidence to suggest that we absorbed them through interbreeding. This latter idea is particular intriguing and DNA sequencing evidence from 2010 suggests that modern non-African humans in Europe and Asia share 1% to 4% of their genes with Neanderthals.

Hobbit or human?

Nicknamed the 'hobbit' because of its small size, *Homo floresiensis* is a recently discovered species said to have lived between 100,000 and 12,000 years ago on the Indonesian island of Flores. In 2003, a female *Homo floresiensis* skeleton was found and dated as approximately 18,000 years old. When alive, she would have been under one meter in height. She could just be a modern human with pathological dwarfism – after all, there were pygmies living on neighbouring islands until 1,400 years ago. Though the fact that this female had a particularly small skull size and therefore brain may keep the debate open a while yet.

The Three Ages

Human prehistory is frequently divided up into three ages: Stone, Bronze and Iron.

All members of the Homo genus, from *habilis* to *sapiens*, existed within a period broadly defined as the 'Stone Age'– which lasted 3 million years or so and only ended between 4500 and 2000 BCE with the advent of metalworking at different times among different human populations.

Because of the enormous length of the Stone Age relative to the metal ages that followed (Bronze and Iron), it has been subdivided into three eras: the Palaeolithic (itself divided into lower, middle and late – characterized by control of fire and use of stone tools); the Mesolithic (first use of advanced technologies including the bow and canoe); and the Neolithic (pottery, general domestication and significant burial/religious site building).

Following the length of the Stone Age, the Bronze Age was a mere blink of an eye. Characterized by the ability to smelt and fashion metals such as copper and bronze to make weapons, utensils and jewellery,

the Bronze Age started at approximately the same time in the most populous regions of the Earth, between 3750 and 3000 BCE in Europe, the Near East, India and China, but later in other areas (800 BCE in Korea, for example) and it ended between 1200 and 600 BCE. Writing is considered to have been invented during this period in Mesopotamia and Ancient Egypt, and the oldest known literary texts date from 2700 to 2600 BCE. Civilizations developed during this period – most notably in Mesopotamia which included Sumer and the Akkadian, Babylonian, and Assyrian empires, all now part of modern-day Iraq.

Next up was the Iron Age, which wasn't just about iron, but also the use of steel. This period started earliest in the Ancient Near East (Anatolia, Cyprus, Egypt, Persia) around 1300 BCE, then Europe and India around 1200 BCE and later in other parts of Asia: China (600 BCE), Korea (400 BCE) and Japan (100 BCE). The Iron Age lasted into the Common Era, ending around 400 CE in Europe and as late as 500 CE in Japan. Significant texts dating from this period include the Indian Vedas, the Hebrew Bible (Old Testament) and the earliest literature from Ancient Greece.

BC or BCE?

Because of its Christian connotations, BC (meaning 'Before Christ') is now often changed to BCE (meaning 'Before Common Era'). AD (Anno Domini, meaning 'In the year of our Lord') is increasingly replaced with the secular CE (meaning 'Common Era'). But whichever way one pitches it, the origin of 'Year One' is unchanged, coinciding with the assigned birth year of Jesus.

Time-Travel Tip
Visit the Grand Canyon!

An astonishing wilderness of rock, the Grand Canyon in Arizona is considered one of the Seven Natural Wonders of the World. A trip to the 277-mile-long canyon is also a trip to 2 billion years of the Earth's geological history – exposed in glorious layer after layer of rock record. Journeying into the perfectly preserved caves and cliff dwellings takes you back to the time when the ancient Pueblo people populated the region around 1200 BCE.

2
Marking Time

Nature's timekeepers

The Earth, the Sun, the stars and the Moon were
following their own cycles and rhythms long before
humans invented the notion of timekeeping. The
planet's rotation, the seasons, the gravitational
effects of the Sun and the Moon, the growth patterns
of trees and plants, are all part of an intricate and
interconnected natural timekeeping that the world
does all by itself.

The Sun and the Moon

Every day as the Earth spins on its axis, the Sun rises
in the East and sets in the West. The planet completes

an annual orbit of the Sun following a pattern that demarks the seasons across the world and dictates the behaviour of all the plant and animal life that draw their nourishment from the Sun. The Sun rises and sets at radically different times depending on where you are. In the tropical areas north and south of the equator, it rises around 6 a.m. and sets at 6 p.m. with reassuring predictability – creating a near-perfect 12-hour day. But at the Earth's poles the lengths of days vary considerably. Sometimes the Sun never sets and sometimes it never rises.

The Sun lights different parts of our Moon each day of the month, though we always see the same 'face' of the Moon. About 29 days and 12 hours elapse between full moons, when the whole face is visible, and from this duration humans created months – though we've since tampered with the length to suit our solar-based calendars, with months now averaging 30.4 days.

The Moon's close proximity to the Earth (some 380,000 km away) creates a significant gravitational pull – strong enough to cause the tides in our oceans. 'High tide' occurs as the Moon passes over the world, pulling the water in its seas up into a hump, which follows around the planet behind the Moon. The

Earth itself is pulled by the Moon, leaving another hump of water on the side away from the Moon that forms the second high tide. So as the Earth rotates there are two high tides every day in coastal areas.

Because it is further away from the Earth, the Sun's gravitational pull is less influential. But when the Sun and Moon align with the Earth, the two gravitational pulls combine to create stronger 'spring' tides. This happens every fourteen days. Because the Moon takes a little more than 24 hours to orbit the Earth, the gap between tides is around 12 hours and 25 minutes.

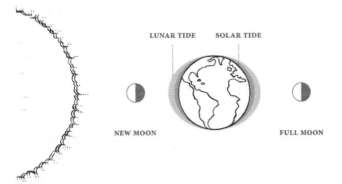

LUNAR TIDE   SOLAR TIDE

NEW MOON   FULL MOON

Spring tides result in higher than average high waters and stronger tidal currents

But the Moon's influence on the sea doesn't end with the tides. It extends to creatures living within it. Oysters open and shut their shells in response to the gravitational pull of the Moon. And apparently the best time to go fishing is at the time of the new moon, when the Moon passes between the Earth, and the Sun is either completely invisible from Earth or visible only as a very narrow crescent.

Once in a blue moon

A 'blue moon' is indeed a rare thing. But it does happen every now and then – twice in every five years or so – when two full moons appear in a 1-month period. If there's a full moon on the first of a month with thirty-one days, then there'll be a second full moon, or blue moon, on the last day of that month.

The phrase can also relate to something a lot more literal, when pollution and dust from volcanoes or fires fill the atmosphere and alter our perception of the moon's colour, making it appear blue.

Tree rings

Every year trees grow during the summer and stop during the cold winter months – creating a new annual growth ring. Thinner rings show that growth was unimpressive in a given year, while the opposite is true of thicker rings. Rings can help us to find out the age of the tree and to gauge the weather conditions in certain years.

The oldest known living tree bears the appropriate name Methuselah. It is a Great Basin bristlecone pine located in Inyo County, California, and in 2013 was estimated to be between 4,845 and 4,846

Methuselah, the oldest known living tree

years old. However, it is believed that an olive tree known as 'The Sisters' in the Batroun district of Lebanon is older still – anything between 6,000 and 6,800 years old.

The so-called Bodhi tree, a sacred fig located in Anuradhapura, Sri Lanka, was planted in 288 BCE and is the oldest living tree to have been grown by humans. According to legend, it was under a sapling from this tree that Buddha became enlightened.

Sun worship

The Sun as a deity appears throughout most of the known ancient religions. Our Neolithic ancestors built great monuments to it to celebrate significant astronomical events. The ancient Egyptians personified it as Ra or Horus, the ruler of the sky, the Earth, and the underworld. The Aztecs had Tonatiuh as their sun god and the leader of heaven, a god that required bloody human sacrifice in return for moving around the world. The Greeks had Helios, a handsome sun god crowned with a shining aureole or halo who drove his chariot of the sun around the world each day.

There are four key dates for our sun-worshipping forebears during the year: the Spring Equinox

Helios, the personification of the Sun in Greek mythology

(20 or 21 March), Summer Solstice (20 or 21 June), Autumn Equinox (22 or 23 September) and Winter Solstice (21 or 22 December). In Europe, two of our most treasured Neolithic monuments, Stonehenge in England and Newgrange in Ireland, were constructed to channel the Sun's rays in a symbolic way on the Summer and Winter Solstices respectively. The Summer Solstice marks the longest day of the year, when the Sun is highest in the sky, the Winter the shortest day, when the Sun is at its lowest. The Equinoxes occur when the centre of the Sun is in the same plane as the equator.

A large proportion of the planet continues to mark the Winter Solstice though many are unaware that that is what they are doing. It is no mistake that Christmas Day is in such close proximity to the Winter Solstice – being a deliberate attempt to co-opt the existing pagan festival. The same can be said of Easter (Spring Equinox) – which is why both celebrations are such a strange blend of Christian and pagan traditions.

Seasons

In the Western world we divide our years up into four distinct seasons (winter, spring, summer and autumn or fall), which tie in with shifts in the weather, and the behaviour of plants and animals in relation to it, and are handily marked by Solstices and Equinoxes.

But such seasons are geographically unique. For example, India recognizes six seasons: hot, rainy or monsoon, autumn, winter, cool season and spring. And in many parts of Africa there are two seasons: dry and rainy. The ancient Egyptians talked of three: flood, winter and summer – each made up of four months. And for a long time the ancient Greeks didn't have an autumn season to go with their spring, summer and winter. Germanic peoples living in the more extreme conditions of Iceland and Scandinavian countries had just two seasons: winter and summer. The words and concepts of spring and autumn were introduced through contact with the Romans.

In Western countries it is widely agreed that the seasons begin in March (spring), June (summer), September (autumn) and December (winter). But in Ireland, for example, the seasons are considered to begin a month earlier, tying in with ancient festivals, most notably Bealtaine on 1 May and Samhain on 1 November – still celebrated with gusto by pagans in Ireland today.

Computus

Computus is the calculation of the date of Easter used by the Roman Catholic Church, since the Middle Ages. In principle, Easter is determined to follow the full moon that chases the Spring Equinox (on 21 March). The earliest possible date for Easter is 22 March, the latest is 25 April.

Months, weeks and days

Months

As previously noted, the time division of 'month' has its origin in the cycle of the Moon, with 29.5 days between full moons. It was only when the notion of the year came into play that the 29.5 number became inconvenient, as it does not tally with the time it takes for the Earth to orbit the Sun (365.25 days) – thus a little extra time had to be added and subtracted to months to make twelve in a year, which is the easiest division.

The names we have for months come from the Romans and all but two date from the eighth century BCE. The New Year then began in March and the first few months were named after gods, for example Mars (now March, god of war), Aprilis (goddess of love), Maia (May, goddess of growth), Juno (June, wife of Jupiter). But then it would seem the calendar inventors changed their minds about this convention towards the end of the year and started using numeric references instead. So the last four months of the first Roman calendar were: Septem (seven), Octo (eight), Novem (nine) and Decem (ten), which we have retained but have skewed the numbers so that their original meaning has been lost (i.e. September is now the ninth month, etc).

January and February were added slightly later at the end of the year, as the Romans thought of winter as a monthless period, and relate to the god Janus (the god of the doorway) and 'februum', meaning purification and linked to a ritual in the old lunar Roman calendar. January came to be seen as the first month of the year in the fifth century BCE.

The names for July and August came later still. The first is for Julius Caesar, who oversaw the implementation of the Julian calendar in the first century BCE, and the second was named after his successor Augustus.

Janus, the Roman god of beginnings, transitions and doorways

Days

The time unit 'day' describes the length of time the Earth takes to rotate on its own axis (around 86,400 seconds or 24 hours). The world's official, 'civil' day runs from midnight to midnight. It is used to determine international time zones and Coordinated Universal Time (the standard we use to set clocks across the globe).

But before we were able to do this fancy counting, days were defined either as the time between sunsets (Ancient Greeks and Babylonians) or sunrises (Ancient Egyptians). In Jewish and Muslim traditions to this day, the day is counted from sunset to sunset.

What a difference a day makes . . .

While a day on Earth is 24 hours, it is considerably longer on our neighbouring planet Venus, where a solar day lasts 116.75 of our Earth days. Mars takes just a little longer than the Earth to rotate at 25 hours, while large planets rotate faster – Saturn and Jupiter's days are just 10 hours long, Uranus's is 18 and Neptune's 19.

Seven-day week

Aggregating days into seven-day weeks is yet another inheritance from the Babylonians and early Jewish civilizations by way of Rome. The days of the week were named for the seven 'classical planets' – those that were visible to early astronomers. They were the Sun (Sunday) and Moon (Monday), and five other planets: Mercury (Wednesday), Venus (Friday), Mars (Tuesday), Jupiter (Thursday) and Saturn (Saturday).

As well as the planet connection, the seven-day week is related to Babylonian and Jewish traditions which endow the seventh day with religious significance. In the Old Testament it was the day God rested after six days creating the Earth, and so Jewish peoples began celebrating a holy day of rest every seventh day. The Babylonians celebrated every seventh day to coincide with the new moon, a quarter of a lunar orbit or 'lunation' (but the timing is a little off so synchronization soon suffered but the seven-day week remained). The ten-day week was favoured by the ancient Chinese and Egyptians as well as in Peru, whereas the Mayans and Aztecs had thirteen days in a week.

Over the years, various nations have tried to tinker with the seven-day-week structure. Between 1929

47

and 1940, the Soviet Union adopted a five-day week. In 1793 in revolutionary France a whole new calendar system was briefly introduced based on the number ten – with ten newly named days per week (more on this later).

In English, our days of the week have been named for a mixture of seven Norse, Anglo-Saxon and Roman 'gods' – which tells the story of various conquerors of Britain over time. The Norse gods are roughly equivalent to the Roman gods (see below). Our neighbours on mainland Europe (e.g. France, Spain, Italy) retain the full suite of Roman gods/planets for their day names.

DAY	GOD	ROMAN GOD
Monday	Moon	Luna (moon)
Tuesday	Tyr (or Tiw in Old English) – Norse god of law, justice, and the sky among other things, including war. He is portrayed as a one-handed man.	Mars – the god of war and all things military.

Wednesday	Woden (Anglo-Saxon equivalent of the Norse god Odin) – the god of war and victory, as well as poets, musicians and seers.	Mercury – the messenger with wings on his sandals. He is also the god of trade, merchants, and travel. He too has a connection with poetry and music.
Thursday	Thor – the Norse god associated with thunder, lightning, storms, oak trees and strength. And famous for his large hammer.	Jove – also known as Jupiter, the king of the gods – and the sky and thunder too.
Friday	Frige – Anglo-Saxon goddess about whom little is known, but it is assumed she is associated with sexuality and fertility.	Venus – the goddess of love, sexuality and fertility whose Greek counterpart is Aphrodite.
Saturday	Saturn – the Roman god associated with wealth, agriculture, liberation, and time, rather than the planet also named after him.	Saturn
Sunday	Sun	Sol (sun)

Calendars

Different cultures devised their own calendars at different times and using diverse markers and measurements, but broadly speaking these calendars fall into two categories: lunar and solar.

The Moon takes 29.53 days to orbit the Earth, so the lunar year runs to just 354.36 days to complete twelve orbits or 'lunations'. Meanwhile, the solar year, based on the Earth's orbit of the Sun, is 365.25 days long.

The Moon is the easier marker of the two, with full moons shining in the sky with reassuring regularity every month. It is therefore likely that lunar calendars were long in use before anyone thought to count a whole year.

Ancient bones

An artefact considered by some to be the earliest physical calendar was found in a cave in the Dordogne Valley in France and is dated as being some 30,000 years old. It is a fragment of bone from an eagle's wing upon which is etched a pattern of notches – the notches appear in groups of 14 or 15 and rows of 29 or 30.

An eagle's bone, found in Abri Blanchard, Dordogne

Could this be a Palaeolithic lunar calendar? Some archaeologists have suggested that the bone might have been used by women to keep track of their menstrual cycles and thus their fertility (though the menstrual cycle is shorter than a lunar month at twenty-eight days). It's an interesting thought. Perhaps this artefact is both a calendar and a contraceptive of sorts. Or neither, of course . . .

West of Kiev in the Ukraine, 20,000-year-old mammoth bones were found with notches that indicated lunar months in periods of four. This has been interpreted as a 'season'.

Time markers

The Neolithic structures at Stonehenge and Newgrange can be considered to be calendars in

51

that they demark and capture time – specifically the exact time of year at the Summer and Winter Solstices. Similar sites in the British Isles that serve as annual markers include Maeshowe on Orkney Island in Scotland and Castlerigg near Keswick in northern England. Such structures are not unique to Europe though. In China, midwinter is captured at Taosi in the Shanxi Province. And in Egypt the temple of Queen Hatshepsut is designed to welcome the midwinter sun.

Another useful annual marker in ancient Egypt was the Nile, which flooded around the same time each year (mid-June, close to the Spring Equinox)

Hatshepsut's Temple, found on the west bank of the Nile

and was used to mark 'New Year'. The flood was considered as one of three seasons that divided the year. The others were growth and harvest. And soon it was calculated that a year from flood to flood was 360 days, subdivided into twelve months of thirty days. Egyptian astronomers noticed that the time of the flood also coincided closely with the day that the sky's brightest star, Sirius, also known as the 'Dog Star', rose in the dawn sky just before the Sun. Using this marker the Egyptians started counting 365 days in the year instead. The resulting calendar became the 'official' calendar of the country used by priests and rulers.

Floods of tears

Ancient Egyptians believed that the Nile flooded every year because of Isis's tears for her dead husband, Osiris.

In Egypt today, the flood event is still celebrated as a two-week annual holiday starting 15 August, known as Wafaa El-Nil. The Coptic Church marks the flood by throwing a martyr's relic into the river in an event known as Esba al-shahid ('The Martyr's Finger').

The Julian calendar

Gaius Julius Caesar observed the usefulness of having an 'official' calendar in Egypt and decided to adopt one for the Roman Empire in the middle of the first century BCE.

The Romans had been counting years and months for quite some time – starting in 735 BCE when the city of Rome was founded by the legendary Romulus (of Romulus and Remus). As previously mentioned, the first Roman calendar had just ten months in it. Then around 700 BCE, another two months were added to the end of the year: January and February.

To bring this old calendar in line with the solar year Caesar employed mathematicians and philosophers to find the most logical system. Through their efforts the Julian calendar was born, naming 25 December as the Winter Solstice (rather than 21 or 22 December – and later co-opted as Christmas Day), but the new calendar was two months behind the solar year – and so for the first year of its existence two months were added to balance things out. This year became known as the 'year of confusion'. Adding to this confusion, Caesar also announced that the year would start in January, rather than March.

The calendar had many teething problems but was soon adopted throughout the Empire – and provided the template for the modern 'Gregorian' calendar (see overleaf). The calendar has twelve months and a leap year every four years (giving February twenty-nine days as it does today).

Sun's day versus Saturn's day

When Constantine the Great became Emperor of Rome at the start of the fourth century CE, he set about adopting Christianity as a way of unifying his crumbling Empire. As part of this project he re-invented the seven-day week. In the Bible it says that while making the Earth, God rested on the seventh day. Constantine therefore decreed that Sunday should be the 'official' day of rest instead of Saturday (Saturn's day). This decision would have meant a fundamental change to the way people lived at the time. Some cultures have never quite adopted the change – with Jews still celebrating the Sabbath on Saturday. Though thanks to the five-day working week we can now pick and choose which day, if either, we decide to practise our faiths on.

The Gregorian calendar

The Julian calendar remained in situ for some 1,600 years before Pope Gregory XIII decided to upgrade it as a rather ambitious pet-project. The main issue with the Julian calendar was that it had wrongly assumed the length of a year was 365.25 days, which is actually 10 minutes and three-quarters too long. Over time this discrepancy pushed the Julian calendar ten days out of sync with the solar year. Gregory's mission was to align the two in a long-lasting way.

First they had to lose ten days. There was much hemming and hawing about how best to do this – not having a leap year for forty years was one popular option. But in the end it was decided to get the pain over with all at once and so the day after Thursday 4 October 1582 was Friday 15 October. While getting people all over Europe to implement this change was an amazing feat, it was not without its problems. People in different countries reputedly felt that they had been robbed of ten days and they wanted them back.

As the edict for change came from the Pope, countries like Protestant Britain were slower to implement the change to the more logical system. In fact, they did not do so until 1752, at which stage eleven days had to be skipped over to align the Gregorian year with the solar year.

Leap year

Occurring once every four years in the Gregorian calendar, leap years are years that have an additional day to help keep the calendar year in sync with the solar year. In a leap year an extra day is added to the month of February, causing a shift or 'leap' in the days. So, in a 'common year' Friday 28 February would have been followed by Sunday 1 March, but in a leap year the first day of March is pushed one day along as 29 February 'leaps' into its place. Leap years are required because the solar year is a little over 365.24 days long.

A person born on 29 February is called a 'leapling' or a 'leaper' and in common years usually celebrate their birthdays on 28 February, though in some places, like Hong Kong for example, a 'leapling's' birthday is legally regarded as 1 March.

What year is it?

While years on the standard Gregorian calendar are measured in relation to the presumed birth year of Jesus, non-Christian communities often benchmark theirs against the birth, death or particularly significant episode in the lives of their own religious leaders. Taking the millennium (2000) as a baseline, here is how some non-Western cultures determine what year it is:

1379: IRAN AND AFGHANISTAN

✱ The Solar Hijri is the official calendar in these two countries – and in many ways is considered more accurate than the Gregorian, though it requires consultation of astronomical charts. To determine the Solar Hijri, you subtract 621 or 622 from the Gregorian year (622 CE was the year the Prophet Muhammad migrated from Mecca to Medina). So the year 2000 equals 1379 in the Solar Hijri.

1421: SAUDI ARABIA AND OTHER
ISLAMIC COUNTRIES

✳ The Islamic calendar is a lunar calendar consisting of either 354 or 355 days (lunar cycles are slightly shorter than the Gregorian month: 29.5 days versus an average of 30.4 days). The calendar is used for religious purposes, for example to determine the appropriate start date for Ramadan. Because of this disparity in the number of days, calculations are a little trickier than for the Solar Hijri, even though both calendars choose 622 CE as their 'Year One'. According to the Islamic calendar, the year 2000 is 1421.

Setting his country back years

Upon seizing control of Libya in 1978, Colonel Muammar Gaddafi reportedly declared that the Islamic calendar should start with the death of the prophet Mohammed in 632 CE, rather than the traditional 622 CE, putting Libya's calendar ten years behind other Muslim countries.

12: JAPAN

✳ While Japan uses the Gregorian calendar for all its official, day-to-day dealings, its year system is rather different and is based on the reign of the country's emperors. So Japan's Emperor Akihito acceded to the throne in 1989, making 2000 CE 'Year 12'.

5760 OR 5761: ISRAEL

✳ The Hebrew calendar used to determine religious days and festivals takes 3761 BCE in the Gregorian calendar system as its start date, one year *before* scriptures say the Earth was created (that happened on Monday 7 October the following year to be precise). To calculate the year 2000 in the Hebrew calendar, one adds either 3760 before Rosh Hashanah (Jewish New Year, usually falling in September or October) or 3761 after Rosh Hashanah. So while the West welcomed the millennium on 1 January 2000 it was 5760 in the Hebrew calendar, and as it prepared to usher in 2001 it was 5761.

Many Christian groups like Creationists and Jehovah's Witnesses still embrace the idea that God created the world in the thirty-eighth century BCE – making our Earth around 6,000 years old.

4637 OR 4697: CHINA

✱ Tradition holds that the Chinese calendar was invented by Emperor Huang-di in the sixty-first year of his reign (2637 BCE). But this is a date with a rather wide margin of error, with others using 2697 BCE, some 60 years later, as a baseline. So depending on what you fancy, it was either 4637 or 4697 in 2000.

5102: INDIA

✱ The Hindu calendar kicks off in 3102 BCE – the year that Krishna is said to have returned to his 'eternal abode'. As the calendar follows the same 365-day and leap-year pattern as the Gregorian, we simply add 2000 to determine that the West's millennium fell in the Hindu year 5102.

1992: ETHIOPIA

✱ Based on ancient Egyptian calendars but similar enough to the Gregorian calendar to allow easy calculation, the Ethiopian calendar celebrates New Year on 29 or 30 August. In terms of its Year One, it is pretty close to the Gregorian with just a seven- to eight-year gap due to a difference of opinion on the year date of the Annunciation of Jesus (when the Angel Gabriel told Mary that she would conceive God's son). The Ethiopians place this event slightly later than the powers-that-be in Rome, making the year 2000 equivalent to 1992 at New Year in Ethiopia and neighbouring Eritrea, and giving them some breathing space before their own Y2K panic kicked in.

Time-Travel Tip
Check out the stars!

On a clear night, take some time to look up at the stars. You won't be looking at them as they are 'now' but how they were when the light left them. That could be millions and millions of years ago. To get the full effect of the night sky, it's best to be in the countryside away from major sources of light pollution like big towns or cities that can block out all but the brightest stars from the sky.

The Sun is around 150 million kilometres away, though this varies, and it takes 8 minutes for its light to reach us. So when you see the Sun, you're seeing it 8 minutes in the past.

3
Keeping Time

Counting the hours

Like many things mathematics and astronomy-related, our time divisions came from ancient Mesopotamia, via the Babylonians and Sumerians, in the third millennium BCE. And also like the number of days in a week – or indeed the existence of weeks at all – these divisions are arbitrary, but exert considerable power over our lives nonetheless.

Why 12 and 60?

The 'sexagesimal' system, which uses 60 as its core number, is used not only for measuring time but for measuring angles and geographic coordinates too, though systems using multiples of ten are used for most other forms of counting and general calculation.

Sixty is a highly composite number – that is one with a useful number of divisors. It has 12 'factors' or ways of dividing into it (1, 2, 3, 4, 5, 6, 10, 12,15, 20, 30, 60), which helps to make fractions involving 60 or multiples of 60 simpler. It is also the smallest number that can be divided by every number from 1 to 6.

Basically, 60 is a great number.

The use of the number 12 as a core number in time and mathematics is thought to come from people counting on their fingers. Specifically, counting the three joint bones in each of their fingers using their thumb as a pointer. There are of course other reasons why 12 has caught on in a big way. There are 12 lunar cycles within a solar year for example, though counting years came long after counting on fingers. The use of 12 as a base for counting is known as the 'duodecimal' system – which was widely used in the earliest civilizations in ancient Egypt, Sumer, India and China.

Hours

Dividing the day by the magic number 12 is largely grounded in the counting preferences of our forebears rather than anything more scientific. Saying that, the ancient Egyptians divided the day according to the rising of 36 'decan' stars, constellations that rise one after the other on the horizon throughout each rotation of the Earth. The rising of each decan signified an hour division (consisting of 40 minutes) and by the 'Middle Kingdom' period (sixteenth to eleventh century BCE) this system had been refined to count 24 decan hours in a day, with 12 for the daytime and 12 for night. But to define the actual length of one of these divisions, a measuring device was required. Enter the very first clocks . . .

Sunlight and shadows

As previously discussed, the Sun is the most useful of nature's clocks. Apart from sunrise and sunset, its easiest point to read is noon, when it is highest in the sky, casting the shortest shadows on the ground. And so this became the most popular point in the day from which to count. It is not known when humans began to use the Sun to calculate the time of day. Basic markers could have been used millennia ago – we just don't know. Nor do we know whether shadows or points of sunlight were first used to tell time, though the earliest sundials we know of indicate that measuring the time of day using shadows was the more common (shadows are longest in the early morning, getting shorter to noon, then longer again towards dusk). The earliest known sundials are from around 1500 BCE, used in ancient Egyptian and Babylonian astronomy.

At its most basic, a sundial is a horizontal or vertical base with a 'gnomon', a thin rod or upright sharp edge that casts shadows onto a surface marked to indicate different times of the day. To give an accurate reading, the sundial must be aligned with the axis of the Earth's rotation and the gnomon must point towards 'true celestial north'. In the northern hemisphere this is indicated by the pole star, Polaris.

A sundial, made up of a horizontal base with a gnomon

The duodecimal markers on sundials were used to measure the length of an hour. This time measurement could then be applied to other time devices such as water clocks, candle clocks and hourglasses – inventions that helped solve the problem of telling the time on a cloudy day and at night.

Water clocks

Water clocks may have been in use as early as 4000 BCE in China, though the only hard evidence we have for them is considerably later, around 1500 BCE in Egypt and Babylon again. Time is measured by the regulated flow of water either into or out of a vessel whose size and flow rate is approximate to a specific time frame.

Examples of basic water clocks include half-coconut shells called *ghati* or *kapala* in India. A small but precise hole was drilled into this simple device that was then placed in a bowl of water. The *ghati* was sized to take 24 minutes to fill with water and sink – with each minute itself being equivalent to 60 seconds apiece. A day was therefore comprised of 60 of these 24-minute hours. The *fenjaan* clocks used in Persia in the fourth century BCE applied the same principle though not the same time measurement.

In Greece, the *clepsydra* or 'water thief' was a water clock constructed from a jar with a hole in the end. When the water ran out, the prescribed amount of time had been measured. To ensure fairness in Athenian courts, a *clepsydra* was used in court cases to fix the amount of time both plaintiffs and defendants were allotted. They are also thought to have been used by prostitutes to measure the time spent with their clients.

To measure time over a longer period required constant maintenance and counting so water clocks became increasingly sophisticated. By the third century BCE, a clock had been invented in

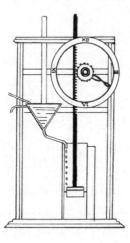

A basic water clock

Greece that used a continuous supply of water and an overflow system – allowing longer periods of time to be measured. Further innovations and mechanisation were slow to develop, though there was a particularly productive period in the Middle East and China between the eighth and eleventh centuries.

Chinese Clock inventor Su Song (1020–1101 CE) created an astronomical water-powered clock, housed in a tower of some 9 metres. The clock featured a celestial globe and panels at the front that opened to display figures holding plaques announcing the time of day.

Another clock, described in an early thirteenth-century text and located in the Umayyad Mosque in the Syrian capital Damascus, split time into 12 equal hours. The clock had dials that indicated the time during the day and night respectively – and copper balls were released to ring the hour.

The hourglass

It is thought that hourglasses or sandglasses were first invented and utilized in Europe in the eighth century. The first evidence of their use is from the fourteenth century, captured in the 1338

fresco *Allegory of Good Government* by Italian artist Ambrogio Lorenzetti. And frequent reference to them is found in ship logbooks from the same time.

An hourglass consists of two connected glass bulbs that allow a regulated trickle of material from the top to the bottom. Once the top bulb is empty, it can be turned and timing begins afresh. They were particularly useful on board ships as they were unaffected by the motion of the sea and the granular material used in hourglass – sand, powdered eggshell or powdered marble – was less susceptible to temperature changes than water-powered clocks. In fact, hourglasses were used to measure time, speed and distance on ships until the eighteenth century.

In the next chapter we'll find out about the first reliable sea clocks or 'marine chronometers' as they're known in the time business – and the extraordinary life of the man who perfected them.

71

Log lines

Another device employed by sailors to measure speed was a 'log line'. This was basically a long piece of rope with a series of evenly spaced knots in it, weighted down at one end with a piece of wood. The wood weight would be cast overboard and sailors would then count the knots in the rope as it was pulled from its coil behind the ship. The counting would happen during a fixed time (usually measured using a small sandglass) and speed would then be calculated in 'knots'.

Burning the candle . . .

Another early 'clock' that was popular across Asia, the Middle East and Europe was the candle clock. In use from at least the early sixth century, but likely earlier than that, the principle was simple – the rate at which the candle burned was used to measure the passage of time.

The candle wax was marked at regular intervals to indicate time periods. Alternatively, the candle could be placed against a marked reflective background and the height of the flame used to indicate the time where the flame lit a marking. Other candles – created to burn within a specified time period – had a nail inside them, which would fall with a clatter once the candle had burned away, announcing the end of the time being measured.

Minutes and seconds

As we have already seen, hours were initially calculated using sundials with divisions of the day based on the duodecimal system (multiples of 12). And we've also learned about the penchant for the number 60 among our forebears from ancient Mesopotamia, from whom we have inherited so much mathematical and astronomical knowledge. So it seems inevitable that the hour as a 1/24 portion of a day would be subdivided by 60 minutes and in turn those 60 minutes each divided into a further 60 short units: seconds.

Before mechanical clocks, measuring how long a second took would have been far from scientific. We can only surmise as to how or if it was done – and it was likely to be as accurate as a child counting in a game of hide and seek. Perhaps they were indicated by steady finger clicks or heartbeats. Coincidentally, the rate of a healthy man's heart comes in around 60 beats per minute throughout his adult life, a woman's just a little over.

Some of the more sophisticated water clocks designed in the High Middle Ages (eleventh, twelfth and thirteenth centuries) were capable of measuring smaller units of time. The early thirteenth-century clock at the Umayyad Mosque in Damascus, for example, also indicated time periods of 5 minutes, as well as its hours. And smaller sandglasses where used to measure shorter periods for a variety of functions.

Fixing seconds

Up to 1960, the second was defined as 1/86,400 of a mean solar day, despite late-nineteenth-century astronomical findings that showed that the mean day is ever-so-slowly lengthening. With the invention of atomic clocks in the 1950s, seconds were captured and defined in exact terms. To give you a sense of the accuracy of these time-devices, there is one in Switzerland, in operation since 2004, that has an uncertainty of 1 second in 30 million years! (More on these on page 138)

But it was with the advent of the first non-water-powered mechanical clocks that minutes and seconds came into their own, and formed the fundamental building blocks for time as we now know it.

The mechanics of clocks

As we've seen, ingenious water clocks with intricate moving parts were in use from at least the eleventh century (though there is anecdotal evidence to suggest they may have been in use over a thousand years earlier in ancient Greece). The breakthrough technology that made them possible is the 'escapement' – an invention that is still used in watches and clocks to this day.

An escapement is a device that transfers energy to the timekeeping element, also known as the 'impulse action', allowing the number of its oscillations to be counted (the 'locking action'). Think of the inner workings of a clock or watch you've seen, with its indented wheels ticking ever onward – it's the escapement that drives this motion and which causes the clock's ticking sound, as the mechanism moves forward and locks, moves forward and locks.

The energy that sets the escapement in continuous motion comes from a coiled spring or suspended weight.

In water clocks, the escapement was designed to tip a container of water over each time it filled up, advancing the clock's wheels with every occurrence. The development of a truly mechanical clock however, required an escapement that could drive the clock's movement using an oscillating weight.

Clocks using this technology began to appear in Europe in the thirteenth and fourteenth centuries. They were necessarily very large and had to be positioned high on a wall or tower because of the sizeable hanging weights required to facilitate continuous motion. Royalty and the ultra-wealthy were alone in being able to afford such devices, and so the majority were commissioned and used by the Church – and housed in monasteries and cathedrals. Their chief function was to call people to prayer.

Medieval timekeepers

The fourteenth century saw the construction of impressive large-scale clocks throughout Europe, chiefly attached to and maintained by cathedrals. These clocks would have required constant

Earliest clock escapement?

Escapements were developed and used as early as the third century BCE by the ancient Greeks to control the flow of water in washstands. There is anecdotal evidence from this time to suggest that this complex technology had already been applied to water clocks. The Greek engineer Philo of Byzantium, creator and user of escapements and author of a treatise on pneumatics, comments that the technology he is using in his washstand is 'similar to that of clocks'.

maintenance and probably resetting due to inaccuracies – but represented incredible strides in timekeeping nonetheless.

Among the many masterpieces produced were Richard of Wallingford's clock at St Albans (1336) and Giovanni de Dondi's in Padua (1348). Though neither clock still exists, we know from detailed descriptions that both had multiple functions. Wallingford's clock had a large dial with astrolabe detailing (a dial device used to locate and predict the positions of heavenly bodies), and an indicator of the level of the tide at London Bridge. Its bells rang on the hour, their number announcing the time. The Paduan clock featured dials showing the time of day, including minutes, the movements of the planets, a calendar of feast days and even an eclipse-prediction hand.

Another lost but reportedly spectacular early clock was at Strasbourg Cathedral. Its most impressive feature was a gilded rooster (a symbol of Jesus) that flapped its mechanical wings and emitted a crowing sound at noon, while three mechanical kings bowed to its splendour. This clock also featured an astrolabe and calendar. Other great fourteenth-century clocks include Wells cathedral's (now at the Science Museum in London and still working), the Gros Horloge at Rouen and the Heinrich von Wick clock in Paris.

Still in operation today and drawing crowds daily is the Orloj in Prague's Old Town Square. Constructed in 1410, this beautiful device combines a mechanical clock, astronomical dial and zodiacal ring, and features many animated figures that are set in motion on the hour. These represent Vanity, Greed/Usury, Death and a Turk (representing sinful pleasure and entertainment). The Twelve Apostles also make an appearance at the doorways above the clock every hour – it's quite a show! It has been repaired and augmented many times over its 600-plus years, and was heavily damaged by German forces during the Second World War.

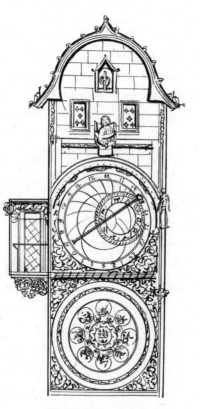

The Orloj in Prague

What's in a name?

The word 'clock' came into usage in the late fourteenth century, replacing the Latin *horologium* (though the practice of clock-making is still known as 'horology'). The reason for this name change relates directly to the earliest common purpose of clocks – that was to call a church's congregation to prayer through the related ringing of a bell. The word 'clock' comes from an earlier word meaning 'bell'. This word is probably Celtic (*clocca* or *clagan*, meaning 'bell') which found its way into medieval Latin, Old French (*cloque*) and Middle Dutch (*clocke*) – all with the same meaning. The use of the word is thought to have been spread throughout Europe by Irish missionaries. The modern Irish word *clog* can mean both 'bell' and 'clock'.

Alive and ticking: the oldest working clock

Older though slightly less beguiling than the Orloj is the clock at Salisbury Cathedral. This clock is thought to date from 1386, making it just six years older than the Wells Cathedral clock mentioned above, which has been dated at 1392.

Some horology conspiracy theorists (yes, they exist!) believe that the Salisbury clock is in fact from a later date, as the construction is quite advanced and similar to clocks made in the sixteenth and seventeenth centuries.

In 1993, a symposium at the Antiquarian Horological Society voted that the Salisbury clock is indeed the older of the two – but around one third of the participants voted against, expressing their belief that the clock is of a much later date. The Salisbury clock has not been in continuous use. In fact, it was only rediscovered after many years' absence in 1928, and was not restored and reinstated until 1956.

Time-Travel Tip
Ignore all clocks!

Our lives are completely dominated by the time systems discussed in this chapter – the constructs of hours, minutes and seconds – captured in devices from the primitive to the sophisticated. Ditch your watch, hide the clock, turn off your mobile and free yourself from this time prison for a few days, and experience time as your prehistoric ancestors would have done – through the movements of the Sun, demarked by dawn, noon and dusk. For the full back-in-time experience (and to safely escape time-keeping devices which are *everywhere*) you're best off taking refuge in a cabin in the woods somewhere and not seeing anybody for the duration. It will likely be a very disorientating experience. Good luck!

4
The Best of Times

Golden age

In this chapter we'll look at some of the great leaps forward in timekeeping during the periods of the Renaissance, the Enlightenment and beyond. During this 'golden age', scientific innovations came thick and fast – from the invention of new mechanisms within clocks and watches to make them ever more accurate, to the physical and philosophical contributions of the likes of Galileo and Newton. We'll journey through some of time's most momentous events, as well as some of its silliest (see 'Cuckoo clocks' on page 116).

We'll also meet some of the timekeeping titans of this golden age. These include John Harrison, inventor of the magnificent maritime measuring machines that revolutionized sea travel, and Abraham-Louis Perrelet, inventor of a self-winding mechanism for pocket watches – technology found in modern wristwatches to this day.

Spring time

The next major development in timekeeping was the invention of the 'mainspring' as the power source for clocks. Replacing weights to drive the escapement, spring-driven clocks appeared in the early fifteenth century. The mechanism works by winding – which twists the spring spiral tight and releases energy as it unwinds over a period of time. The earliest existing spring-driven clock belonged to the Duke of Burgundy (now part of modern France) and dates from 1430.

Later in the fifteenth century, clocks which indicated minutes and seconds began to appear – though none indicating seconds have survived (the earliest example is from 1560). Before that, most clocks had just one hand, with the face split into four sections of 15 minutes.

The advent of the mainspring led to a boom in clock and watchmaking, especially in the German cities of Nuremberg and Augsburg. As the technology became more affordable, demand soared.

Small timekeepers were very fashionable in the mid-sixteenth century and would have been worn ornamentally on a chain around the neck or fastened to clothing. They would have required regular winding – twice or more times a day.

The well heeled of Nuremberg who wanted to stand out from the crowd commissioned all manner of unusual and eye-catchingly shaped watches – representing animal forms, flowers, insects and skulls. The face of watches was exposed, though many had lids to protect the hands. Glass over the face only began to be used as standard in the early seventeenth century.

In 1510, the German master clockmaker Peter Henlein (1485–1542) created the 'Nuremberg Egg', one of the first-known watches. Henlein is often credited with 'inventing' the watch, though the Nuremberg of his time was bursting with talented clockmakers dead set on turning out ever-smaller and more intricate timekeepers for their fashionable clientele.

Indeed, the business was so competitive that it turned violent. In September 1504, Henlein

The Nuremberg Egg

was involved in a brawl with fellow locksmith/clockmaker George Glaser, in which his rival was killed. The details are sketchy but we do know that Henlein fled to a local Franciscan monastery were he sought sanctuary for four years. By 1509 he was back in favour and was appointed the master of Nuremberg's locksmith guild.

King Phillip and the clockwork monk

When King Phillip II (1527–1598) of Spain's son and heir Charles suffered a serious brain injury his father was naturally distraught. And when Charles miraculously recovered, the king believed it could only be the work of God favouring his family and answering their prayers. Phillip vowed that he would thereafter honour God with continuous prayers. But being king is a busy job. So, instead of doing the praying himself, Phillip commissioned an automaton (aka robot) to be made to do his praying for him.

Using the latest in clockwork technology, the automaton was made of wood and iron, standing at 15 inches tall. It was driven by a key-wound spring and could walk in a square, strike its chest with one hand, and raise and lower a wooden cross and rosary with the other. When operated, its head nodded, turned, rolled its eyes and opened and closed its mouth in rickety 'prayer'.

Four hundred years later, this perpetual prayer machine is still in good working order and lives in the Smithsonian Institution in Washington, DC.

Huguenot horologists

The centre of European clockmaking activity moved from Germany to Switzerland in the early sixteenth century – following a highly skilled group called the Huguenots. Originally from France, the Huguenots were followers of John Calvin and members of the Protestant Reformed Church during the sixteenth and seventeenth centuries. Highly critical of the Catholic Church, they were victims of intense persecution. During the St Bartholomew's Day massacre in 1572, up to 30,000 Huguenots were slaughtered by Catholics in Paris, sparking similar attacks in provincial towns and cities in the weeks that followed. This brutality was officially sanctioned – the perpetrators were pardoned for their actions against the Huguenots.

It is estimated that as many as 500,000 Huguenots fled to Protestant countries including England, Denmark, the Netherlands and North America. But it was in Switzerland that they established themselves as master clockmakers and gave that country the leading edge, which it is still famous for to this day. King Henry VIII welcomed the Huguenots into England. In fact, he personally brought a group of clockmakers over from France to attend to the clocks in his palaces.

Big swing

The horological innovations continued apace, though the one I'm about to describe took quite some time to get from idea to practical application.

The story goes that the Italian polymath Galileo Galilei (1564–1642), known variously as the 'father of observational astronomy', the 'father of modern physics', and the 'father of modern science', was in the Cathedral or Duomo in Pisa in 1582. The young student was reportedly wiling away the service observing the swinging motion of a large bronze lamp in a draught. Galileo timed the swings against the beats of his pulse and found that regardless of how wide the swings of the lamp, the time between them was always the same – nine or ten pulses. He got home and tried a number of experiments and found that the length of the rope used to create the pendulum motion affected the swing rate – the longer the rope, the longer the swing time.

Galileo applied his findings to the creation of a portable pulse meter, which he used in his medical work. The usefulness of the device was soon recognized by the medical establishment. Later in life, Galileo discovered that the pendulum could be applied to clocks and drew up plans for the first pendulum clock in 1637, but he never built it. His son Vincenzio started work on the clock in 1649, but died before he completed it.

It was another polymath, Christiaan Huygens from the Netherlands, who finally brought Galileo's vision to life in 1656 when he constructed the very first pendulum clock – which was accurate to less than 1 minute per day, highly accurate at the time. In 1675, Huygens also invented the spiral balance spring for the balance wheel of pocket watches, making them significantly more accurate.

While Huygens's pendulum clock was certainly revolutionary, the type of escapement it used, known as 'verge', gave too wide a pendulum swing

A basic pendulum clock

and its accuracy was compromised. The invention of the 'anchor' escapement around 1670 reduced the pendulum's swing significantly, and the shorter the swing the greater the accuracy of the clock. The anchor quickly became the standard escapement used in pendulum clocks, though who actually invented it is unknown. Clockmakers Joseph Knibb and William Clement, as well as the scientist Robert Hooke, are variously credited with its creation.

Grandfather clocks

The preferred swing range of pendulums was narrowed further still and soon the seconds pendulum came to be favoured, swinging once per second. The long narrow clocks built around these pendulums were first crafted by Englishman William Clement around 1680 – and became known as grandfather clocks. Minute hands were also introduced as standard around 1690.

The name 'grandfather clock' is thought to have come from a popular song from the 1870s called 'My Grandfather's Clock', written by abolitionist Henry Clay Work. The eponymous clock resided in the George Hotel in Yorkshire, England, and was renowned as a very accurate timekeeper, that is until one of its two owners passed away and it began to mysteriously lose time. When the second owner died, it reportedly stopped working altogether.

The metronome

The invention of the pendulum led to the creation of the first prototype metronome in 1696. Designed by French musical theorist Etienne Loulié, the device had an adjustable pendulum that could be set to different speeds, but it did not make a sound nor have an escapement to keep the pendulum in motion.

Another hundred years-plus would pass before the metronome proper was invented, this time in the Netherlands in 1814 by Dietrich Nikolaus Winkel. Though Winkel invented it, another man called Johann Maelzel developed, patented and started manufacturing metronomes in 1816 under the name 'Maelzel's Metronome' lest there be any confusion. It was designed as a tool for musicians to keep a steady tempo – at various speeds. The tempo is measured in beats per minute (bpm) ranging from 40 to 208 bpm.

Five rather interesting facts about Galileo

So as well as coming up with the pendulum, Galileo Galilei made significant contributions to the fields of physics, maths, astronomy and philosophy.

1. As a medical student and practitioner he invented not only a pendulum-based pulse meter but a 'thermoscope', a forerunner to the thermometer.

2. He was an instructor at the art school Accademia delle Arti del Disegno in Florence where he taught perspective and *chiaroscuro* (the lighting effect used by Caravaggio and Rembrandt).

3. He made improvements to the technology of the telescope, equipping himself to confirm the phases of Venus and to discover the four largest lunar satellites of Jupiter, named the Galilean moons in his honour. He also studied sunspots and the Milky Way, which had previously been largely dismissed.

4. He championed the Copernican theory that the Sun rather than the Earth was at the centre of the 'universe'. This met with opposition from both his astronomical peers and the Church. He was accused of heresy, tried and forced to spend the last fifteen years of his life under house arrest for daring to challenge the establishment. In 1939, Pope Pius XII described him as among the 'most audacious heroes of research' and eventually, in 1992, Pope John Paul II issued an official apology for how Galileo was treated, on behalf of the Roman Catholic Church.

5. Among his other achievements, he invented a military compass that allowed greater accuracy in cannon use, created a compound microscope, described an experiment for measuring the speed of light, and put forward a principle of relativity that provided the basics for Newton's laws of motion and Einstein's special theory of relativity!

Pocket watches

Outside of church clock towers, time and timekeepers were very much the preserve of the moneyed – and their practical uses were often secondary to their material expression of wealth and fashion. This was especially true of the pocket watch. In fact, it is thought that the pocket watch evolved to complement a burgeoning fashion of the late seventeenth century – the waistcoat.

Actually, it's not entirely true to call the waistcoat a 'fashion'. The generously bewigged King Charles II of England, Scotland and Ireland introduced the waistcoat as a part of 'correct' dress during the Restoration of the British monarchy (his father was executed by the government of Oliver Cromwell in 1649). The diarist Samuel Pepys wrote in October 1666 that 'the King hath yesterday in council declared his resolution of setting a fashion for clothes which he will never alter. It will be a vest, I know not well how.'

The pocket watch evolved, with its close-fitting lid and rounded edges, to slip neatly into the pockets of such a vest or waistcoat.

Sir Isaac Newton

It's hard to know where to place Sir Isaac – so great was his contribution across so many of the areas featured in this book. His book *Philosophiae Naturalis Principia Mathematica* (Mathematical Principles of Natural Philosophy, first published in 1687), set out Newton's laws of motion and universal gravitation – paving the way for the study of physics thereafter.

Among his other achievements were his astronomical calculations, which cemented the belief that the Sun is the centre of our 'universe' rather than the Earth (which went without the punishment meted out to Galileo for the same). He built a reflecting telescope, studied the speed of light and contributed to the development of calculus.

Newton distinguished between 'absolute' time and 'relative' time. In his conception, time was 'not liable to any change' – it exists without us, is independent and absolute, and it progresses at a consistent pace throughout the universe. According to Newton, people can only perceive 'relative' time, which we measure through perceivable objects in motion like the Moon or Sun – or indeed clocks. Through these movements we create our sense of the passage of time.

But despite his apparent scientific rationalism, in later life Newton dedicated a great deal of his time to alchemy (trying to turn base metal into gold) and the study of biblical chronology. In fact, through his work on the latter, Newton estimated that the world would end no earlier than 2060 – though he would give no firm prediction as to when it would actually end. Instead, he stated: 'This I mention not to assert when the time of the end shall be, but to put a stop to the rash conjectures of fanciful men who are frequently predicting the time of the end, and by doing so bring the sacred prophesies into discredit as often as their predictions fail.'

The end of time

Most world religions, the Abrahamic (Judaism, Christianity and Islam) and non-Abrahamic, have specific teachings on the 'end time'. Across belief systems, the end time is usually characterized by a period of tribulation, redemption and/or rebirth, ushering in a new era where life is eternal.

Renowned theologian Hippolytus of Rome and others predicted that Jesus would return in the year 500 CE and usher in the end time with his second coming. Following this non-event, others including

97

Pope Sylvester II (946–1003) predicted the end on 1 January 1000. The anticipation of this millennial apocalypse brought thousands of pilgrims to Jerusalem, as the ground zero of the Christian end time.

When this too failed to occur, other Christians decided the end would happen on the 1,000-year anniversary of the death, rather than the birth, of Jesus – 1033. Determined to keep the anniversary theme alive, 2000 was the next obvious year to focus on. The possibility of Jesus reappearing to do battle with the Antichrist tended to be overshadowed by grim apocalyptic visions of nuclear holocausts, asteroid strikes – and of course the much anticipated technological disaster of Y2K.

Recent years have seen a glut of predictions. Fear of the Large Hadron Collider brought apocalyptic visions of the planet being devoured by black holes. The 'Rapture' was due to happen on 21 May 2011 according to US Christian radio host Harold Camping. He predicted that on that date around three per cent of the world's population would ascend into heaven and the rest of us would die horribly with the Earth five months later on 21 October. Camping had previously stated that the Rapture would occur in September 1994. Rather than throw his rather battered hat in the ring a third time, Camping announced that his attempt to date the end of time was 'sinful' and that his predicting days were numbered.

Many people got stirred up with talk of the Mayan apocalypse – based on a very subjective reading of a stone inscription. The world was due to end on 21 December 2012 and again defied predictors.

Undiminished, apocalyptic predictions for our third millennium CE abound among Christian, Muslim and Jewish theologians. Scientists are a little more generous with the end date though, giving planet Earth at least another 5 billion years or so – at which point it will likely be swallowed by the Sun. Though before that, as the Sun grows hotter, life on the planet may become impossible in a mere 1 billion years. In turn, the 'Big Rip' theory suggests that the entire universe will eventually be torn apart by its continuous expansion in around 22 billion years' time.

Setting the time

Up until the 1670s, nobody outside of the maritime world really cared about a fixed concept of time – and that remained the case until quite a few years later. Time was localized not synchronized. In the absence of mass communications or basic infrastructure, it really didn't matter what time it was in the next town or city – only the ringing of your own church bell counted.

But for mariners, knowing the time was essential. It gave them control over navigation (more to follow on this), and also of the tides – knowing when the tide would be high or low according to their tide tables was vital for organizing sailing times. In London, the hub of maritime life in Stuart England, Greenwich on the River Thames was selected as the place where clocks would be set before voyages.

The Royal Observatory, Greenwich

In 1675, the restored King Charles II of England (he of the waistcoats decree) established the Observatory at Greenwich as the place where his Astronomer Royal would 'apply himself with the most exact care and diligence to the rectifying of the tables of the motions of the heavens, and the places of the fixed stars, so as to find out the so much desired longitude of places for the perfecting of the art of navigation'.

The Observatory was designed and built by Sir Christopher Wren, the man who rebuilt St Paul's

Cathedral and countless churches destroyed in the Great Fire of London, as well as designing the massive Royal Greenwich Hospital for Seamen at the bottom of the hill upon which the observatory stands. The Observatory has the further distinction of being the first purpose-built scientific research facility in Britain, housing the finest equipment and telescopes in the land.

But while this worthy work of mapping the heavens for the benefit of English seafaring was going on, the Observatory's foremost use was a time collection point for mariners disembarking from the docks of Deptford and Greenwich. In the 20-foot-high Octagon room were two clocks created by Thomas Tompion (see 'Timekeeping titans', on page

112), each with an enormous pendulum measuring 3.96 metres and giving time to an unparalleled accuracy of 2 seconds per day. Before setting sail, clocks and watches would be set at Greenwich. But despite careful maintenance, accuracy could not be safeguarded for long at sea – and we'll soon see how one master clockmaker, John Harrison, dedicated his life to creating the perfect marine timekeeper.

In 1833, to save mariners trudging up the steep hill, a time ball was installed on top of the Observatory. This bright red orb was, and indeed still is, raised just before 1 p.m. every day and drops exactly on the hour, so sailors in situ on the Thames could set their marine chronometers accordingly. A few years later in 1855, the Shepherd Gate clock was mounted on the wall outside the Observatory. It has a 24-hour dial and is an early electric slave clock – driven by electric pulses transmitted from the master clock inside the main building. It is thought that this clock was the first to display 'Greenwich Mean Time' to the public.

From 5 February 1924, the British Broadcasting Corporation began transmitting a time signal direct from Greenwich. The 'pips' as they were known, were intended to help people to set their watches

Flower clock

How well the skilful gardener drew
Of flow'rs and herbs this dial new;
Where from above the milder sun
Does through a fragrant zodiac run;
And, as it works, th' industrious bee
Computes its time as well as we.
How could such sweet and wholesome hours
Be reckoned but with herbs and flow'rs!

Andrew Marvell, The Garden, 1678

Many years after Andrew Marvell wrote the above lines, botanist Carolus Linnaeus wrote about the idea of a flower clock in the 1751 publication *Philosophia Botanica*. Subsequently a number of botanic gardens attempted to plant flowers as he suggested – that is, so flowers bloomed and closed in sequence, demarking different times of the day. Sow thistle for example, typically opens at 5 a.m. and closes at 12 p.m. Hawkweed opens at 1 a.m. and closes at 3 p.m. The latest closing flower recommended by Linnaeus was the Day-lily at 7 to 8 p.m. Seasonal and weather changes make this a rather tricky clock to maintain.

and clocks to the correct time. There were six pips in total, the last one longer than the others and announcing the exact moment of the start of the next hour. Because radio waves travel at the speed of light, the pips could be transmitted to the far side of the world and still give a time reading accurate to around a tenth of a second.

The pips are no longer broadcast from Greenwich, but from the National Physical Laboratory in Teddington, Surrey, which uses Coordinated Universal Time (UTC) – the successor of GMT – for its reading.

Maritime time and the longitude debacle

The 'Age of Sail' between the sixteenth and nineteenth centuries saw a worldwide revolution in trade and human movement around the planet. But as we've read, sailing was a precarious adventure, not least because of the difficulties in determining longitude – the location of a place on Earth east or west of a north-south prime meridian line.

Seafarers used calculations based on astronomical maps and live readings of the stars to try to determine their location at sea – but with the result that they often missed their end destinations by a considerable margin, or in worst cases experienced shipwrecks and lost lives. So great was the problem that in 1714 a competition was announced to find the best solution to the longitude problem, with the British Parliament offering a considerable prize of £20,000 (approximately £2.9 million in today's money).

Self-educated carpenter and watchmaker John Harrison from Yorkshire took up the challenge to invent a sea clock capable of keeping time in the harshest conditions and thus aiding a simpler, time-based calculation of longitude. In the process he pitted himself against the astronomical establishment. His chief competitor was Nevil Maskelyne, an astronomer with strong support on the Board of Longitude working on a 'Method of Lunar Distances' for calculating longitude.

Harrison invented five masterpieces of maritime timekeeping over a 40-year period – each one breaking new ground in horology. He started with the large and beautifully ornate H1 and finished with the deceptively simple H4 and H5 (oversized pocket watches which could withstand the shocks of sea travel). When H4 was tested on a transatlantic journey to Jamaica, it was just 5 seconds slow.

When the ship returned, Harrison expected to be awarded his £20,000. He was wrong. The Board of Longitude stated that this accuracy could be luck and requested further trials.

On H4's second journey, this time to Barbados, it was accurate to within just 39 seconds. Also on this second voyage was Nevil Maskelyne, testing his Method of Lunar Distances for measuring longitude, which was accurate to within 30 miles – an impressive result, but still not as strong a performance as Harrison's H4. Plus Maskelyne's calculations required considerable time and effort, unlike the sea clock.

Harrison's H4

Again the Board of Longitude said that H4's accuracy was a matter of luck and required that it undergo further testing by the Astronomer Royal, the newly appointed Nevil Maskelyne. Maskelyne unsurprisingly returned a very negative report on the sea clocks' performance and scuppered Harrison's chance of claiming the prize.

Though he felt 'extremely ill used' by the establishment, the dogged Harrison began work on H5 and enlisted the support of King George III – who himself tested the clock, reported on its incredible accuracy and advised Harrison to petition parliament for the full prize. At the age of 80, Harrison eventually received £8,750 of the longitude prize money but he never received the official award, nor did anyone else for that matter. He died three years later in 1776 at the age of eighty-three – bucking another time trend of the era by living so long. By the early nineteenth century, the use of sea clocks was the norm for establishing longitude in maritime travel.

Prime meridian

For the purpose of global navigation, the prime meridian is the agreed point of 0° longitude which encircles the Earth. This notional line divides the planet into eastern and western hemispheres, just as the equator divides it north and south. However, unlike the equator, the position of the prime meridian is arbitrary. As a consequence, many countries have tried to claim that the invisible 0° line should pass through their little patch.

The first recorded meridian line is found on Ptolemy's 'world' map of 150 CE, though the idea of a prime line of longitude dates to the third century BCE. Ptolemy's map consists of about a quarter of our globe – stretching west to east from the Canary Islands in the Atlantic off Spain as far as China. And under the Arctic circle in the north to the top half of Africa in the south.

Ptolemy's world map, 150 CE

The meridian line on this map passes through El Hierro, one of the Canary Islands, as it was the westernmost body of land known at the time. This map and location of the prime meridian were influential in cartography right up to the late fifteenth century, until explorers such as Christopher Columbus started to rapidly increase the size of the known world.

The focal point of the line moved a little south and west to Cape Verde off the coast of Africa on the advice of Columbus, becoming known as the Tordesillas line, after a treaty between Spain and Portugal to settle territorial disputes over newly discovered land. The line fluctuated between Cape Verde and the Canaries for another 200 years until the early eighteenth century, when the British went hell-for-leather trying to solve the longitude problem and assigned Greenwich as the point the prime meridian passes through. With so much nautical information and guidance pouring forth from Britain – the Greenwich meridian soon became the norm. In 1884, an International Meridian Convention held in Washington, DC officially agreed that Greenwich was indeed the site of the prime meridian, though the French continued to use Paris until 1911. Greenwich Mean Time was also established as the standard time from which the rest of the world should measure its time of day.

The site of the prime meridian in Greenwich is a major tourist draw, with lines of people queuing to get their picture taken straddling the eastern and western hemispheres. At night the observatory shines a green laser beam into the sky to proudly demark the line that determines the degrees and measurements on every contemporary map of the world.

The revolutionary power of ten

Not a country to be told to fall in with either standard time or a standard calendar, France attempted to break with both during the years following its revolution. The French Republican Calendar was used for about twelve years from 1793 and was adopted as part of France's bid to embrace decimalization (using 10 instead of 12 as a fundamental unit), as well as to divorce the calendar from religious associations.

The government abandoned the Christian system for years, dating the new calendar from the birth of the Republic (Year One being 1792), and though it continued to split the year into twelve months, these were divided into three ten-day weeks called *décades*. Keeping the decimal theme going, these days were split into 10 hours, each made up of 100 decimal minutes of 100 decimal seconds. So the new hour was more than twice as long as the old one of 60 minutes of 60 seconds. Even the minutes got longer – they were equivalent to 86.4 seconds rather than 60, though the seconds themselves were shorter at 0.864 of a conventional second.

Clocks were created to report decimal time, but their makers were not inundated with orders. Decimal time was only mandatory for two years and abandoned completely in 1805, as ironically it proved a complete waste of time.

The names for the 'new' months were vivid and evocative, relating to nature and the weather. The autumn months, for example, were Vendémiaire (Grape Harvest), Brumaire (Fog) and Frimaire (Frost).

The ten days of the week were somewhat more functional: Primidi (first day), Duodi (second day), Tridi (third day) . . . ending with Décadi (tenth day) which was a day of rest equivalent to Sunday.

Rather than have saints' days like the once-dominant Catholic Church, the French adopted a system of assigning an animal, plant/food, mineral or tool to each day of the year. The twenty-eighth day of Vendémiaire (22 September to 21 October), for example, is the day of the tomato, and the fifth day of Frimaire (21 November to 20 December) is the day of the pig.

Timekeeping titans

The seventeenth and eighteenth centuries saw so many great innovations in timekeeping and the creation of such a vast array of intricate, beautiful and sometimes just plain weird timekeepers that doing them justice would require another book. However, there are a few heroes of horology who must be mentioned.

Thomas Tompion (1639–1713)

Referred to as the 'Father of English Clockmaking', Thomas Tompion created the first two 'regulator' clocks for the Observatory at Greenwich, used by the very first Astronomer Royal, Sir John Flamsteed. Accurate to within two seconds a day (the most accurate in the world at the time), these clocks could run for a full year without rewinding – and they continued to run while being rewound. They were used to literally 'keep' time at the observatory – providing the time for all other clocks and watches in use there and for seafarers.

Year Zero

Inspired by the French Republican Calendar, the Cambodian despot and head of the Khmer Rouge, Pol Pot, declared 1975 to be Year Zero to mark the occasion he took control of Phnom Penh, the largest city in the country.

But Pol Pot was very literal-minded with his changes to time in the country – and he was determined to alter not only the calendar but the epoch in which Cambodians lived. He sought to de-industrialize the country, levelling the societal playing field by effectively making everyone a member of an uneducated, peasant class. The country's history was to be erased – and so intellectuals, teachers and artists, who might keep the cultural memory alive, were targeted for persecution.

During the four years of Pol Pot's rule, approximately 2 million people lost their lives as a consequence of political executions and forced labour.

Tompion employed a number of skilled French and Dutch Huguenots (who we know were reputed for their horological talents) in his workshop, which may account for the consistently high quality of the timekeepers he produced. Tompion's workshop built about 5,500 watches and 650 clocks during his career. He also created a serial numbering system for his spring and long-case clocks, perhaps the first for manufactured goods.

George Graham was Tompion's most famous protégée and ultimately his business partner. Graham invented the 'Graham dead-beat escapement' in around 1715, which developed on the escapement first made by Tompion in 1675 for the Greenwich clocks. His support for John Harrison was also invaluable – he loaned him £200 so he could start work on his first marine chronometer, H1.

Julien Le Roy (1686–1759)

This master craftsman belonged to the fifth generation of a family of clockmakers and made his first clock at the age of just thirteen. He moved to Paris from his hometown of Tours a year later and rose through the ranks of guilds, the Société des Arts, and ultimately became the official clockmaker or *Horloger Ordinaire du Roi* to King Louis XV in 1739.

Le Roy made many mechanical innovations, including a special repeating mechanism that greatly improved the precision of watches and clocks. He made one for Louis XV that is thought to be the first to allow the owner to remove the clock face to see the intricate inner workings.

During his professional life, Le Roy and his workshop produced some 3,500 watches – around 100 per year – while other workshops would have produced somewhere in the region of thirty to fifty timepieces per year. Examples of his work are housed in the Louvre in Paris and the Victoria and Albert Museum in London.

Continuing the dominance of this clockmaking dynasty, Le Roy's son Pierre (1717–1785) is responsible for three major innovations in horology that paved the way for the modern precision clock and marine chronometer, the latter inspired by the work of Englishman John Harrison. These are the detent escapement, the temperature-compensated balance and the isochronous balance spring.

Cuckoo clocks

For a pretty daft invention, the cuckoo clock has some impressive forebears. The Greek mathematician Ctesibius fashioned a water-driven automaton of an owl for his second-century BCE clock. It whistled and moved at certain times. Then in 797 CE, Harun al-Rashid of Bagdad gave Charlemagne a clock from which sprang a mechanical bird to sound the hours. And the renowned fourteenth-century clock in Strasbourg Cathedral featured a gilded rooster, which flapped its mechanical wings and emitted a crowing sound at noon each day.

There were a few clocks featuring mechanical cuckoos in the seventeenth century, but the eighteenth saw a veritable flurry of them coming out of the Black Forest region of south-west

Germany – but we don't know who started the trend or why. They became increasingly more elaborate and intricate with time. So much so that the *Guinness Book of Records* has a category for the World's Largest Cuckoo Clock, which at the time of writing resides in Sugarcreek, Ohio, and is 23 feet tall and 24 feet wide, featuring a five-piece band, a couple dancing a polka and, of course, a large cuckoo singing in the half-hours.

Cuckoo clocks have a metaphorical association with madness – and there are a remarkable number of them which are said to be haunted. Yes, haunted. I've come across accounts of one that supposedly starts by itself and goes straight to the right time unaided, and another that produces a ghostly apparition when it chimes midnight. Some of the mechanical birds are thought to have wicked intentions – with the spirit of a real bird trapped within. Given that cuckoos are such nasty nest-stealers by nature, it's scarcely surprising.

Abraham-Louis Perrelet (1729–1826)

In the 1770s, this ingenious Swiss horologist invented a self-winding mechanism for pocket watches. The mechanism works using the oscillating up-and-down motion of a weight as its owner walks – operating on the same 'automatic' principle as the modern wristwatch. A test conducted by the Geneva Society of Arts concluded in 1777 that 15 minutes walking would keep a Perrelet watch ticking for eight full days. Another of his inventions was the 'pedometer' – a device that measured steps and distance while walking, and now a rather popular, though usually digital, item for avid walkers and runners.

The Perrelet brand is still producing luxury timepieces in Switzerland, and claiming in its promotional tagline to be the 'Inventor of the Automatic Watch'.

Time-Travel Tip
Past-life regression

Put your scepticism aside and book yourself an appointment with a hypnotist. As they lull you into a trance you may have the potential to travel back in time from the comfort of their leather couch.

Practitioners of past-life regression believe that you can recover and relive memories from your own past – forgotten or repressed – or go even further back into previous incarnations, lived in different physical bodies long deceased.

Even if you don't believe what you find yourself saying, you'll learn something of the vividness of your imagination and the gems you have stored in the recesses of your wonderful, complex mind.

5
Modern Times

Standard time

We saw in the previous chapter how Greenwich became the centre of timekeeping and the site of 0° longitude. But despite these advances it would take another 150 years for the notion of 'standard' time to take hold – and that was down to the arrival of the railway.

A brief history of rail travel

The idea of rail travel, that is pulling goods along a purpose-built surface, goes right back to ancient Greece in the sixth century BCE. Spanning 6 kilometres of grooves cut into limestone, the Diolkos 'wagonway' was used to transport goods (in trucks pushed by slaves) for over 600 years. Railways using tracks or grooves appeared from the fourteenth century and by the sixteenth century narrow-gauge railways with wooden rails were common in European mines.

Britain led the way in the development of more ambitious railway lines. By the seventeenth century, wooden wagonways were in common use for transporting coal from mines to canals, and horse-drawn railways sprang up throughout the eighteenth century. The Industrial Revolution saw the invention of the steam engine and in 1825 an engineer called George Stephenson built the 'Locomotion' for the Stockton and Darlington Railway, north-east England, which was the first public steam railway in the world. He followed this with the intercity railway line from Liverpool to Manchester, which opened in 1830.

The steam locomotives built by Stephenson were soon in use throughout Britain, the US and Europe. By the early 1850s, Britain had over 7,000 miles of rail track.

In America, building railways was a much more laborious undertaking, given the sheer scale of the country, but they were critical to pioneering

An early steam locomotive

businessmen who wanted to open up access to the west. Keeping a keen eye on developments in Britain, America opened its first small-scale railways in the 1830s and 1840s, but it was the period between 1850 and 1890 that saw rapid expansion of the railroads and America becoming home to one third of the total track mileage on the planet. The first transcontinental railroad was aptly completed in 1869 following the civil war, connecting the country together for the first time.

Train time

With the surge in railway building across Europe and America, it became pretty obvious that a standard time was needed for services to run efficiently.

Greenwich Mean Time was first officially used by the railway system in Britain in 11 December 1847 – with every train having its own portable chronometer set to GMT. And to facilitate the precision of 'railway time', as it became known, the Royal Observatory in Greenwich began to transmit time signals by telegraph in August 1852.

However, it took another thirty years-plus for standard time to replace local time on the US railway system. The different railway companies in America set their own time standards. The leading standards were New York time, Pennsylvania time, Chicago time, Jefferson City (Missouri) time and San Francisco time – and with so many competing 'local' times in use, things got rather confusing.

In October 1883, the heads of all the US and Canadian railway companies met in Chicago and agreed to adopt a four-time-zone standard (five time zones are now in use). On 18 November 1883, all the railways readjusted their clocks as per their relevant time zones, though Standard Time was not enacted into law in the US until 1918.

The railways literally carried standard time around these countries and in just a few years all time was set to match it – except the time kept by the British Post Office that is, which continued to be 'London time' rather than GMT until 1872. GMT became the legal time in Britain in 1880.

Time zones

There are twenty-four time zones in use across the world, which use the notional lines of longitude to define their boundaries, with each one taking Greenwich Mean Time (GMT) as its reference or 'offset' time. Every 15 degrees of longitude adds or subtracts an hour to/from GMT depending on whether it's going west (minus) or east (plus) – with 360° of longitude ultimately adding up to 24 hours.

Some nations and territories are more flexible in terms of how they interpret time zones and longitudinal boundaries. Our two most populous countries in the world, India and China, apply a single time zone to their vast expanses (more on time in China shortly). India also uses half-hour deviations, along with Newfoundland, Iran, Afghanistan, Venezuela, Burma, the Marquesas, and parts of Australia. Other nations and provinces, including Nepal and the Chatham Islands, use quarter-hour time deviations.

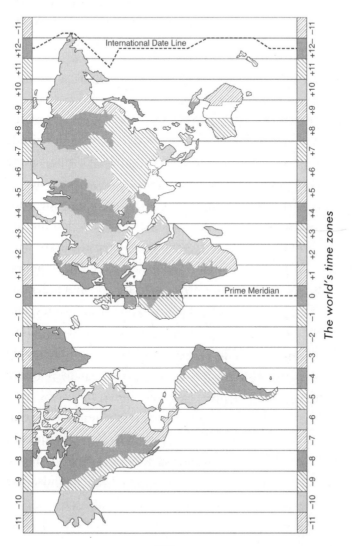

The world's time zones

Coordinated Universal Time

Commonly known as UTC, this system is largely a continuation of Greenwich Mean Time, with the terms GMT and UTC being used interchangeably with much the same meaning – to laymen at least. The introduction of UTC was led by the International Astronomical Union, which called for a more stable and accurate time standard that took into consideration the Earth's natural 'wobble' – and the fact that the Earth's rotation is slowing ever so slightly due to the drag of the tides.

The abbreviation of Coordinated Universal Time to UTC rather than CUT may seem a little odd. In French it would be *Temps Universel Coordonné*, or TUC, so as a compromise with France it was agreed that Universal Time Coordinated or *Universel Temps Coordonné* would be used and abbreviated to UTC, even though it is most commonly referred to as Coordinated Universal Time.

Many would be unaware of this, but GMT ceased to be the world's 'official' time standard in 1972. UTC is held by a number of atomic clocks, some 260 in total, in forty-nine different locations around the world (more on atomic clocks soon). The master of all these clocks is at the US Naval Observatory in

Washington, DC. Though it's unlikely to captivate the imagination as the centre of time quite like leafy, lovely historic Greenwich (I've lived there – I'm rather biased).

All the time in China

Technically, the colossal country that is modern China straddles five time zones, but it uses only one. Officially the entire country is 8 hours ahead of GMT (or UTC). The decision to have one time zone instead of five came in 1949 when the Chinese Civil War saw the end of the Republic of China and the start of the People's Republic of China. With this new communist era, the country was unified under one time, known as Beijing Time, though out of necessity local time continued to be unofficially used in western parts of the country that are up to two and a half hours behind Beijing Time.

Mecca Time

While Greenwich is the largely agreed upon site of the prime meridian for secular, administrative purposes, there are a number of other meridians dotted around the world – the most interesting among them relating to holy sites.

The Great Pyramid of Giza, the largest and oldest of the mighty pyramids of Egypt, for example, was an obvious and popular choice for the line to pass through up until the latter part of the nineteenth century. The Church of the Holy Sepulchre in Jerusalem was a popular meridian point for devout Christians but again it didn't catch on globally. But much more recently Mecca, which is the centre of the Muslim world, has been proposed as a new and appropriate site for the prime meridian. Time at the Mecca meridian is UTC+02:39:18.2.

The idea that Mecca should become the focal point of the prime meridian came in 2008, when Muslim clerics met in Doha, Qatar, at a special conference titled: 'Mecca: the Centre of the Earth, Theory and Practice'. Then the world's largest clock (more on this later) ticked into life in Mecca at the start of Ramadan in August 2010 – displaying Mecca Time – well sort of. In fact, it was ultimately set to display Arabia Standard Time, which takes its lead from the meridian at Greenwich rather than leading with its own.

Daylight saving time

Daylight saving time is observed largely in the northern hemisphere: Europe, Canada and America, along with a couple of African and Latin American countries, New Zealand and part of south-east Australia – though many more countries observed it in the past, including Russia, China and India. The basic principle is adding an hour in spring and subtracting that hour back in autumn to make our days a little bit longer and brighter in the evening – in summer at least.

Daylight saving time was first introduced in the early twentieth century during the First World War by Germany and its allies as a way of saving coal and other energy sources. Their enemies – Britain, France, et al – decided this was a good idea and followed suit, as did many of the neutral countries on the peripheries of the war. By 1918, Russia and the US had adopted daylight saving time too.

Many countries abandoned daylight saving time in the years following the war with the exception of the UK, France, Ireland and Canada – though other countries dipped in and out – and with the coming of the Second World War it was widely adopted once more. The 1970s energy crisis saw a spike in popularity for pushing clocks forward in springtime to reduce the amount of fuel used for electric lighting.

From pocket to wrist

In the last chapter we saw how the trend in portable pocket watches sprang from the seventeenth-century fashion for waistcoats – and in popular culture the image of the well-heeled gent, with a gold chain across his midriff, is familiar from that time right up to the early twentieth century. But with ever-increasing numbers of watches being produced, and continuing advances in timekeeping technology, the pocket watch went from being an item sported only by the wealthy to a much more ubiquitous one. Functional as well as fashionable, the pocket watch had to evolve to keep the market fresh and so, in the early twentieth century, along came the latest must-have item: the wristwatch.

A clock of birds

The Kaluli people of Papua New Guinea still live their lives by the 'bird clock'. The early morning calls of certain birds tell the people to get up and the afternoon calls of others tell them to go home, ensuring that people are safely back in their villages while visibility is still good.

Of course, the crowing of cocks still heralds the start of the working day in rural areas in the West, though they are likely supplemented with an alarm clock just in case. Like humans, chickens have a 'circadian' cycle – with biological processes relating closely to the 24-hour day. Light cycles exert influence over their heart, brain and liver functions and, in male chickens, testosterone, which relates to their crowing behaviour.

Chickens have been known to shift their roosting and crowing patterns in response to changes in light intensity caused by the changing of the seasons or living on higher ground. In a mountainous area of northern India it was found that cocks started 2 to 3 hours before sunrise. The intervals between their crows increased the closer it came to sunrise, despite the fact that the Sun was not visible to them.

Wristwatch No. 1

The story goes that in 1904 the aviator Alberto Santos-Dumont asked his friend, the French watchmaker Louis Cartier, to design a watch for him that he could easily refer to while flying – the pocket watch being an inconvenient item to consult mid-air. And so Cartier developed the first wristwatch. Well, technically it wasn't the first – the Swiss watchmaker Patek Philippe developed the 'lady's bracelet watch' in the 1860s as a stylish piece of timekeeping jewellery for the woman about town – but Cartier's wristwatch design caught on in a way that PP and others could only dream of. And the main advantage it had was war.

As well as seeing the first widespread use of daylight saving hours, the First World War also saw a surge in popularity for the wristwatch – a much more convenient item for an officer to wear on the battlefield.

Luxury brands

Like the pocket watch before it, the wristwatch started life as an exclusive item, worn by middle- and upper-class men because of its initial expense. The 'inventor' of the men's wristwatch Louis Cartier marketed his first 'Santos' watch to the great and good in 1911. He followed that in 1912 with two models that are still on sale today – the 'Baignoire' and 'Tortue' – and because the war created such demand, he also released the rather macho-sounding and still popular 'Tank' in 1917.

Cartier remains a luxury brand to this day – selling high-end watches and jewellery. Visit the Cartier website and you'll find the least expensive watch on for a tidy £1,600, while the price of the most expensive is undisclosed – you have to request it. But the uppermost price on display is an impressive £50,000 for a diamond-encrusted, white-gold affair.

A browse through many of the online catalogues of the most luxurious of the luxury watch brands yields the same silence on price. Most of the most exclusive companies are Swiss and pretty old – TAG Heuer, Vacheron Constantin, Breitling, IWC, Zenith, Audemars Piguet, Girard Perregaux, Blancpain, Patek Philippe, Piaget, etc. It would clearly be vulgar to put a price on watches so special.

Steel and plain old gold are the cheapest materials relatively speaking. But white, yellow and pink gold see the prices soar, and they go rather stellar when you throw in some precious stones, platinum, titanium or palladium.

The most expensive timepieces

At the time of writing, the Chopard 201-Carat Watch is the most expensive watch on the market, coming in at a cool $25 million. And it is also one of the most hideous-looking things around. A blur of gaudy precious stones, with a tiny watch face tucked away in there somewhere, it's impossible to know how you'd even wear the thing. Amusingly, the second most expensive is a pocket watch made by Patek Phillipe in 1933, priced at $11 million. Indeed, vintage Patek Phillipe watches tend to fetch millions at auction without fail – take that, Mr Cartier!

Back in 1999, a Thomas Tompion clock from 1705 fetched over $2 million at Sotheby's and would likely make much more than that if re-auctioned today. But the current clock record is held by a French design by Abraham-Louis Bréguet. Built in 1795, this rare Sympathique clock is currently valued at $6.8 million.

Clever clocks

Advances in physics in the twentieth century revolutionized timekeeping and clocks became very, very clever indeed. It's all a bit much for my unscientific mind, but in the following pages I've done my best to outline the chief advances that have changed the way we tell time for ever.

Piezoelectricity

Some solid materials – crystals, ceramics, biological materials like bone – accumulate and store an electric charge. Known as piezoelectricity (from the Greek 'to squeeze' – as squeezing releases the energy), it was first discovered and demonstrated in 1880 by brothers Jacques and Pierre Curie (husband of Marie). They revealed how an electrical charge could be generated when mechanical force was applied to crystals (including quartz), sugar cane and Rochelle salt.

Piezoelectricity was subsequently used in sonar devices developed during the First World War, including an ultrasonic submarine detector, phonograph cartridges, telephony devices and aviation radios, among other innovative new technologies.

But more importantly for us, piezoelectricity powers the quartz crystal oscillator that is the driving force in most modern wristwatches.

Quartz timekeepers

Quartz crystals have been used in both clocks and watches since the 1960s. When electric pulses are applied to the crystal it vibrates – and these vibrations can be fine-tuned to any desired frequency. For clocks and watches, the crystal is cut into the shape of a tiny tuning fork and manipulated until it vibrates to a frequently of 32,768Hz – equivalent to a 1-second pulse. This is all terribly precise and revolutionized timekeeping.

The first quartz clock was developed in 1927 and the National Bureau of Standards in the US used quartz time as the time standard for the whole country from 1929 until the 1960s. The first quartz wristwatch came onto the market in time for Christmas in 1969 and cost the same as a small car. Despite the high price, Seiko's Astron model sold well and, with research and development, quartz watches were soon affordable for the majority.

Omega Speedmaster and NASA

In 2013, the popular men's body-spray brand Lynx unveiled a new advertising campaign during the coveted US Super Bowl ad break spot. The ad features a statuesque and fearless lifeguard undertaking a daring rescue of a damsel being distressed by a shark. After he's beaten up the shark and returned the young lady to land, the pair share a tender moment until she spies a man approaching wearing an astronaut suit. She abandons the lifeguard and races into the arms of the astronaut as the tagline 'Nothing Beats an Astronaut' appears on screen.

But the truth is nothing beats the extraordinary advertising coup of the Omega Speedmaster watch back in the late 1960s. First, it was endorsed by NASA as spaceflight-ready, then it was worn during the first American 'spacewalk' on the Gemini 4 mission in 1965 (when astronaut Edward H White floated around in space outside his ship for 20 minutes), and *then* it was worn by none other than Neil Armstrong when he took those first steps on the Moon.

Atomic clocks

While quartz oscillators still proliferate in clocks and wristwatches, such devices are no longer used as the source of standard time. Earlier in the chapter I mentioned that Coordinated Universal Time is held and maintained by a number of atomic clocks, some 260 in total, in forty-nine different locations around the world, with the master of all these located at the US Naval Observatory in Washington DC. But what is an atomic clock?

These clocks are accurate to within 1 second every 30 million years. Born out of particle physics in the 1930s and 1940s, atomic clocks use minute vibrations emitted by electrons in atoms to calculate time. There are 9,192,631,770 atomic vibrations in every second. The first accurate atomic clock was invented in 1949 by the American physicist Isidor Rabi (1898–1988).

Atomic clocks are used to control the wave frequency of television broadcasts, and in global navigation satellite systems – from which the GPS in your car or mobile phone draws its data.

Quantum clocks

A close relation of the atomic clocks, quantum clocks bring aluminium and beryllium ions together in an electromagnetic trap and cool them to near absolute zero temperatures. Now, I can't pretend to know what that does, except that vibrations are involved, but I do know that it makes quantum clocks even more accurate than the atomic clocks which are the current keepers of standard time – more than thirty-seven times more accurate apparently. The most accurate of these ultra accurate clocks was built in February 2010 by the clever people at the US National Institute of Standards and Technology. It uses a single aluminium atom and is expected to lose just 1 second in 3.7 billion years. Though how this will be tracked is another issue entirely.

Why-oh-why-2K?

Reading about an invention as complex and precise as the atomic clock, accurate to within 1 second in 30 million years, it's hard to grasp the relative silliness of Y2K. Also known as 'the millennium bug', Y2K was going to be the end of us all – because we hadn't programmed our computer technology to cope with a change of date from 31 December 1999 to 1 January 2000, from the twentieth into the twenty-first century. With most computers using a two-numeral system to represent the year date i.e. 99 rather than 1999, it was thought that the change to 00 would bring about confusion and chaos in our global systems. So systems were upgraded in a hurry toward the end of the 1990s, but that didn't stop a media frenzy predicting the end of modern civilization.

People started stockpiling food and saying prayers to protect themselves against this impending technological apocalypse. But when the date came, the apocalypse was nowhere to be seen. Though there were some computer failures, the exact number isn't known, because, well, it's a rather embarrassing thing to admit. Saying that, there were a couple of scary occurrences as a result of the 'bug'. In Japan, radiation monitoring equipment failed and at a nuclear plant an alarm sounded just after midnight causing panic. A significantly less scary thing happened in Australia, where ticket validation machines on buses failed in two states. And in America some slot machines in Delaware gave up the ghost.

Bearing in mind my initial comment about atomic clocks, the US Naval Observatory that runs the master atomic clock for UTC gave the wrong date on its website on 1 January 2000 – posting the year as 19100 instead – as did France's national weather forecasting service. Proofing against Y2K cost over $300 billion worldwide.

For the record

So far this book has sought to capture the history of time and timekeeping, and to report on some of the phenomenal advances in technologies and some of the silliest too. But before we take a leap into the future in the next chapter, it's time to take stock of where we are now and take a peak at some mind-boggling records of the day.

The shortest time ever measured

Back in 2004, scientists claimed to have measured the shortest interval of time ever: 100 attoseconds. An attosecond is one quintillionth of a second. One attosecond is to a second what a second is to around 31.71 billion years – more than twice the commonly held age of the universe. The 100 that have been measured, if stretched so that they lasted 1 second, would last 300 million years on the same scale. It boggles the mind!

We're unlikely to hear much about attoseconds in our day-to-day lives, but milliseconds, microseconds and nanoseconds are already here and will be increasingly important in the future.

First off, let's consider the second itself. In Chapter 3 we touched on the ancient origins of the duodecimal (12) and sexagesimal (60) counting systems and how the idea of a second was born of that (24 hours of 60 minutes, each subdivided into smaller units of 60 – aka seconds). Up to 1960 the second was defined as 1/86,400 of a mean solar day, but now it is measured by atomic clocks and defined by atomic vibrations, so 1 second equals 9,192,631,770 vibrations.

When you're dealing with numbers that big, there is plenty of scope to drill down into smaller and smaller parts. A millisecond is a mere one-thousandth of a second, or one beat of a midge's wings (a housefly's takes about 3 milliseconds). Milliseconds are handy for measuring computer activities, which tend to operate much faster than the human mind. For example, computer monitor response times tend to be between 2 and 5 milliseconds.

A microsecond is one millionth of a second, or a thousandth of a millisecond. It takes the human eye around 350,000 microseconds to blink. These and nanoseconds (a mind-boggling one billionth of a second) are used to measure light speeds and sound frequencies. There are also picoseconds (one trillionth of a second) and femtoseconds (one quadrillionth of a second) which are used to measure things like the vibrations of atoms in molecules.

The longest-running clock

The longest-running clock, that we know of at least, is the 'Beverly Clock', which lives in the reception of the Department of Physics at the University of Otago, New Zealand. Constructed in 1864 by

Arthur Beverly, the clock has never been manually wound, ever. Instead, its mechanism is driven by perpetual motion caused by variations in atmospheric pressure and changes in daily temperatures. The temperature variations either

cause the air in a 1-cubic-foot airtight box to expand or contract, pushing on the clock's internal diaphragm. A variation of 6°C over the course of a day will create enough pressure to lift a one-pound weight by one inch and drive the clock's mechanism onward.

Now, the clock may never have been wound, but it has actually stopped a few times. On occasion the mechanism has needed to be cleaned or it has failed and needed to be repaired, or at other times the temperature variations have not been sufficient enough to power the clock.

At the University of Oxford in England lives the Oxford Electric Bell, or 'Clarendon Dry Pile', which is an experimental electric bell that has been continuously ringing since 1840 – well, almost continuously. Thankfully for the other occupants of the building that houses the bell, its ringing is inaudible behind two layers of glass.

The biggest clocks

The biggest clock in the world, in terms of the size of its visible workings, is the same clock discussed earlier as the keeper of 'Mecca Time'. This gargantuan clock sits atop the Abraj Al Bait Towers in Mecca, Saudi Arabia. Its face has a diameter of 43 metres. The clock tower that houses it is the tallest in the world (and the second tallest building in the world), and the building that houses the tower has the world's largest floor space. It'll be a while until anyone trumps this place.

Other notably enormous clock faces are the Cevahir Mall clock in Istanbul (36 metres) and the Duquesne Brewing Company Clock in Pittsburgh (18 metres). The famous Big Ben in London is a mere 6.9 metres in diameter (you could fit six of them on the Mecca clock face with room to spare).

The littlest clock

What qualifies as a clock these days is up for discussion. Clocks are embedded in most of our technological devices and invisible apart from that digital display in the lower right-hand corner of our computer monitor or the screen of our mobile phone.

The current record holder for the 'smallest atomic clock' was constructed at the National Institute for Standards and Technology (NIST) in Colorado in the US. It was unveiled in 2004 and is the size of a grain of rice and is accurate to 1 second in 3,000 years – so considerably less accurate than the most accurate atomic clock in Switzerland, which is accurate to 1 second in every 30 million years. But still, a clock the size of a grain of rice and that accurate is pretty impressive.

The longest-running time experiment

Back in 1927, Thomas Parnell of the University of Queensland, Australia, commenced what is now the longest-running time test: the pitch drop funnel experiment.

Pitch is a petroleum product, an elastic polymer, with a tough rocklike appearance. When heated up pitch becomes highly malleable and is used for waterproofing boats. At room temperature pitch feels solid – even brittle – and can easily be shattered with a blow from a hammer. If heated and left to its own devices it takes years for it to change and move.

Parnell was curious about this material and wanted to demonstrate the fluidity and high viscosity of pitch, so he heated a batch of it and poured it into a funnel. He sealed off the funnel, allowed the pitch three years to settle, trimmed off the end of the funnel and waited for it to drip. Eight years passed. And then the first drip fell from the pitch, in a blink of an eye, in December 1938 – eleven years after the experiment commenced.

Since that time the pitch has slowly dripped out of the funnel. At the time of writing, over eighty years since the experiment began, the ninth drop is only just forming for its all-too-brief journey out of the funnel. The time it takes for the drips to come out is inconsistent. For example, the sixth drop fell in April 1979, 8.7 years after the previous one, but it then took 9.3 years for the seventh to come out in July 1988 and then 12.3 for the next one in November 2000.

Images of the funnel can show a large droplet of pitch hanging tantalizingly from its mouth – looking ready to drop at any moment – but still with years to go. When it does drop, it takes just an eighth of a second. The experiment's current custodian is John Maidstone. He's been watching it since January 1961 – and he has never seen it happen. He has missed this grand event five times. In 1988, he missed it while making a cup of tea. In November 2000, he set up a camera to monitor it as he was away in London, only to find that the camera had failed and the drop was not captured. Maidstone described it as one of the saddest moments of his life.

The fact is nobody has ever seen it drip. But next time they definitely will – whether in real time or captured on camera. Not willing to risk missing it again, Maidstone now has three cameras continually focussed on the pitch to capture the moment it drops. And, rather sadly, people across the world are watching it live online. You can too at the University of Queenland's School of Mathematics and Physics website.

If you find you develop a taste for such things as a consequence, I recommend www.watching-grass-grow.com – the web address tells its own story.

The longest and shortest lives

Relative to other creatures with which we share the world, humans live a long time, though life expectancy varies widely depending on where you are born (see page 206).

According to botanist and ecologist Ghillean Prance, 'The shortest biography is said to be that of the mayfly: Born. Eat. Sex. Die. Pausing neither to eat nor to court, mayflies emerge from the nymph stage with all the food they need for their adult life, and mate in flight. Typically they live only one or two days.' It is worth noting that the immature part of the mayfly's life, the 'naiad' or 'nymph' stage, can last up to a year, before its all too brief adulthood.

The brief stages of a mayfly's life

Insects tend to dominate the 'shortest living' category. Among mammals, the house mouse probably has the shortest innings, with those living to four years being in the geriatric class. Among fish the mosquitofish is an OAP by two, and among birds hummingbirds are way over the hill if they make it to seven or eight.

In the animal kingdom, the Asian elephant has been observed to live as long as eighty-six years, while the oldest living bird is the macaw, which can live up to 100 years in captivity. The lizard-like tuatara reptile of New Zealand can live up to 200 years, while the Japanese koi fish can live more than 200 years in the right conditions. One such fish, called Hanako (meaning 'flower maid') was reportedly 226 years old upon her death in 1977. Greenland sharks, native to the North Atlantic, are believed to live to around 200, and the slow-moving Galápagos tortoise can keep going till around 190 according to current data – and there are many examples of tortoises living in excess of 150 years. The Bowhead whale is also thought to live to around 200 if life is relatively incident free.

The longest-living known creatures are molluscs in the bivalvia category (whose bodies live in shells of two hinged parts). One quahog clam, affectionately known as Ming (comparing its great age to that of the Chinese Ming Dynasty), was believed to be 405 to 410 years old when discovered (and killed) off the Icelandic coast in 2007. Its age was judged by the annual growth rings on its shell – it is unknown how long this creature may have continued to live on the ocean floor had it not been 'discovered'.

There are sponges near Antarctica which are thought to be at least 10,000 years old and black coral in the ocean off New Zealand that may be 4,000-plus years old.

We should consider the speed at which life is lived and experienced by these different creatures. The tortoise lives its long life with a glacial slowness that befits its age, while the tiny hummingbird can beat its wings as often as 90 times per second when hovering, while some short-lived flies like midges can beat theirs more than 1,000 times per second. So another way of looking at it is that hummingbirds and flies pack as much into their little existences as tortoises do – just a hell of a lot faster.

The oldest old folks

According to the Guinness Book of Records, Jeanne Calment of France lived the longest life on record – dying in 1997 at the tender age of 122 years, 164 days. Prior to Calment the record was held by Japanese centenarian Shigechiyo Izumi – though it turned out that Mr Izumi's record could not be verified and he may have been a mere 105 at death rather than 120. Typically, Izumi's longevity was not hampered by the fact that he put away a daily dose of booze and took up smoking at 70. Japanese woman Misao Okawa who is 115 at the time of writing, took possession of the Guinness World Record as the oldest living person in June 2013, when her Japanese compatriot Jiroemon Kimura expired at the tender age of 116.

Time-Travel Tip
Dash across the
International Date Line

This imaginary line around the Earth passes through the middle of the Pacific Ocean, following the similarly imaginary 180° longitude line. Well, it doesn't so much follow it as zigzag in proximity to it as it journeys from the North to the South Pole. On either side of this imaginary line it is a different calendar day. A traveller crossing the International Date Line going east has to subtract a day or 24 hours, heading west they add a day. So criss-crossing the International Date Line allows us to travel back and forward in time by 24 hours!

In Jules Verne's *Around the World in Eighty Days* (1873), Phileas Fogg believes that he has lost his famous wager to complete his eighty-day journey by the evening of Saturday 21 December 1872. Disappointed, a little humiliated and believing it to be Sunday 22 December, Fogg realizes, just in the nick of time, that he forgot the Date Line in his calculations and that he did in fact complete the journey in seventy-nine days, and dashes to claim his prize. Good old-fashioned time travel in action.

6
Future Time

Real time

In August 2011, I was living in Greenwich, London, the home of time. Between the 6th and 11th of that month, multiple riots broke out all over London and elsewhere in England. Out the window of my flat on Blackheath Hill I could see a helicopter hovering over nearby Lewisham, where clashes with the police and looting were taking place. Meanwhile I had the television tuned to the live news on the BBC where the images captured by the aforementioned helicopter were being broadcast, and in my hand I held a smartphone upon which I was following live reports on Twitter from the ground in Lewisham.

The events were happening, being reported, accessed and processed in 'real time'. In real time, events are captured and transmitted at the same rate that the audience experiences them. And for a generation of young technology users, real time is, well, real time. Information is instantaneous, as is our interaction with it. This is the new normal for now. But real time is set to get a whole lot faster.

Instant messaging

The biggest change in mass communications since the advent of postal systems was the invention of the telegraph. The first electrical telegraphs were sent in Germany in the 1830s and could travel a distance of around 1 kilometre. The subsequent flurry of activity on both sides of the Atlantic in developing the technology and laying the requisite cable meant speedy advances in telegraphy in the 1830s and '40s, most notably by Sir William Fothergill Cooke and Charles Wheatstone in the UK, and Samuel F. B. Morse, of Morse code fame, in the US.

By the 1850s the first commercial telegrams had been sent the 750 miles between New York and Chicago – taking a mere quarter of a second to travel that distance (that's 11 million miles per hour).

The first people to use it could barely fathom that such a thing was possible. By the 1860s a transatlantic telegraph cable was in operation, and by the 1870s Britain was wired to its faraway colony India. In 1902, the telegraph system spanning the Pacific was complete and the world was fully encircled by wires, relaying and receiving information over vast distances at previously inconceivable speeds. And then it went wireless.

Using radio technology, pioneer Albert Turpain sent and received his first Morse code radio signal in France in 1895. It only travelled 25 metres, but was a considerable achievement. The following year an Italian called Guglielmo Marconi sent his first radio signal a full 6 kilometres. Marconi took his technology to Britain and the rest is history. In 1901, the first wireless transmission, the letter S, was sent across the Atlantic.

Concurrent with these developments, other inventors were working on transmitting not just signals, but human voices through wires. The electric telephone was invented in the 1870s and the first commercial services were established in New Haven, Connecticut, and London in 1878 and 1879 respectively. Telephone exchanges were established

in every major city in the US by the middle of the 1880s, but it wasn't until 1915 that the first US coast-to-coast, long-distance telephone call was placed – from New York to San Francisco. And it would be another twelve years before human speech could be carried across the Atlantic, when in 1927 radio was used to transmit voices back and forth.

It's hard to imagine how profound an impact these new forms of mass communications had on everyday life and people's perceptions of the world they lived in – and of speed and time. Where once a letter would take weeks and even months to bring news of its writer across the world, messages could now be relayed in moments. But once these things became the 'new normal' people came to expect them, and to expect faster, better ways of sending and receiving information at that.

The innovations continued thick and fast. In the UK, the BBC put out its first radio transmissions in 1922, and by 1925 some 80 per cent of the country was being reached through regional and relay stations. Also in 1925, the Scottish inventor John Logie Baird demonstrated the transmission of moving pictures (just silhouettes at that time) at Selfridges department store in London. Two years later the cathode ray tube was invented and the BBC started its first experimental broadcasts in 1932, with an expanded service launching from Alexandra Palace

The transmitter at
Alexandra Palace

on a high hill in North London in 1936 – a world first.

Flash forward to today, through the innovations and inventions of colour television, videotelephony, satellite phones, radio and television and all the advances in computer technology. On a single device we can now access full-colour, high-definition television or listen to radio in 'real time', play games, read the newspaper, receive video calls, send email, and communicate with friends, family and the wider world instantly through myriad social channels, take photos or video footage – and tell the time – all while walking down the street. And it's only been possible to do all of these things together since the late 2000s. Now that's time speeding up.

If we want to slow it down again, we can always watch the pitch drop experiment livestreamed from Australia on this same device. It's due to drop any day now . . .

The speed of money

The average person's electronic interaction with their money is a speedy affair. Banks communicate with each other instantly, relaying the information necessary for us to make a quick financial transaction almost anywhere in the world, or to complete multiple transactions from the comfort of our desks through online banking. But this speed and ease is laughable next to the lightening-fast activity on today's stock markets.

On the stock market, where countless goods and financial products are traded – the tangible and intangible – 50 to 70 per cent of all the trades are executed by an algorithm with no human input. And buying and selling is conducted in milliseconds.

A 'high-frequency' electronic trader might do 1,000 trades in a minute, but for every trade conducted there are numerous uncompleted transactions that disappear into the ether. These high-high speed computers test the market, sending out buy-and-sell orders and when another computer connects with an order, all the others that weren't taken up are cancelled.

Further computer programs have been designed to identify and defy other similar trading algorithms. These jump into the market, push the price up and sell to other algorithms, making huge sums in seconds.

In the New York Stock Exchange there is a room that is 20,000 square feet (about three football fields) filled with row upon row of servers, around 10,000, owned by various financial institutions and each analysing 'the market' and trading. This is all done without any human involvement and significantly faster than any human can think, let alone act.

Information wars

In the highly competitive world of stocks and shares, the speed at which information travels from, say, the commodities market (basic goods) in Chicago to the equities market (stocks in companies) in New York is critical to closing deals faster than the other guy. Every millisecond counts. Fibre-optic cables allow information to travel between these markets in 15 milliseconds. But traders want that information even faster. This demand started a race to get the straightest and therefore fastest fibre-optic line from Chicago to New York to shave a millisecond or two off the speed at which information travelled, providing much-valued time to the ultra-fast trading computers.

The speed of light through air is even faster than fibre optics. So to capitalize on this, towers are now being constructed to beam information between the trading centres in an estimated 8 milliseconds. One day soon these transmissions may be conducted in microseconds, or possibly even nanoseconds.

Speed dating

So I hope we've established that the way we live and the way we interact with each other is accelerating. And as a consequence time is becoming ever more precious. In this fast, frenetic, multi-tasking world we're streamlining everything, including how we find our partners. Enter speed dating. Rather than spending all that time trying to meet 'the one', you can now go to one place and meet a number of 'ones', talk to them for short intervals of between 3 and 8 minutes, and if a connection is made before the bell rings, note it, pass it on to the organizers and let them tell you whether your potential beloved returns your affections. Cue wedding bells.

Faster and faster and faster

Information travels far faster than humans can. But humans can travel pretty fast. The fastest footspeed record currently belongs to the appropriately named Usain Bolt, who made it up to 27.79 mph (44.72 kph) during a 100-metre sprint in 2009. Bolt completed the 100m race in 9.58 seconds, beating his own previous world record of 9.69 seconds. Humans are still considerably slower than other animals. Cheetahs are the fastest creatures on Earth and, in 2012, a Cheetah called Sarah created a new world record by running 100 metres in 5.95 seconds, reaching a top speed of 61 mph (98 kph). Saying that, even a domestic cat could outrun Usain Bolt – reaching recorded speeds of 30 mph (48 kph).

Travelling faster and reducing the time spent getting from A to B has been a key human endeavour. Next, we'll look at some of the fastest modes of transport ever invented and consider the future of how we travel at ever greater speeds.

Planes, trains and automobiles

The Wright brothers' first successful engine-powered flight in 1903 reached a whopping 6.8 mph (10.9 kph) speed. By 1905 their speed record was up to 37.85 mph (60.23 kph). Today's airspeed record for a manned flight was set by the Lockheed SR-71 Blackbird in July 1976. This bird got up to 2,193.2 mph (3,529.6 kph).

The fastest commercially available car is the Bugatti Veyron Super Sport, which can go from zero to 60 mph in 2.4 seconds and reach a top speed of 267 mph (431.07 kph). You can get your own for just $2.4 million. There are, however, no roads upon Earth on which you can legally drive at that speed – the highest legal speed limit is a mere 150 kph in Italy (followed by 140 kph in Poland, Bulgaria and the United Arab Emirates). There is no speed limit on the German autobahn, though 130 kph is recommended. Travelling 300 kph faster in a Bugatti would likely not go down too well.

To give a sense of just how fast we've speeded up over time, the first commercial automobile powered by petrol, designed by German Karl Benz (of Mercedes-Benz), hit the road in 1888 with a maximum speed of just 16 kph. The fastest train on the planet is currently the CRH380A in China, which has a top speed of 302 mph – making it the fastest legal way to travel by land.

164

Breaking the sound barrier

The speed of sound is 343.2 metres per second (or around 768 mph). The sound barrier was first encountered during the Second World War when aircraft started to see the effects of compressibility – an aerodynamic effect that struck their crafts, impeding further acceleration. Hitting the sound barrier in an unsuitable craft creates loud cracks or 'sonic booms'. Design changes to aircraft, making them more aerodynamic, allowed them to break through this barrier and increase acceleration. The sound barrier was officially broken by American Chuck Yeager in 1947 flying an XP-86 Sabre.

The first time a land vehicle broke the sound barrier was just one year later in 1948, when an unmanned rocket sled reached 1,019 mph (1,640 kph) before jumping off its rails. The first manned vehicle was driven by Briton Andy Green in 1997, when his vehicle, the Thrust SSC (supersonic car), achieved a top speed of 763 mph (1,228 kph).

In October 2012, Austrian Felix Baumgarter became the first skydiver to travel faster than the speed of sound, reaching a maximum velocity of 833.9 mph (1,342 kph). To achieve this he jumped from a balloon floating 24 miles (39,045 metres) above New Mexico (and way above the stratosphere) – and in doing so also broke the record for the highest-ever freefall. To put that in perspective, your average Boeing 747 reaches a maximum height of 13,000 metres and Mount Everest peaks at just 8,848 metres. Baumgarter's fall to Earth took just over 9 minutes, with only the last 2,526 metres negotiated by parachute.

Rocket speed

In May 1969, the *Apollo 10* space rocket set off on a dry-run mission, testing all the procedures required for landing on the Moon – without actually landing on the Moon. That was done by *Apollo 11* in July of the same year. During that mission, *Apollo 10* is thought to have reached the highest speeds ever attained by a manned vehicle – 24,791 mph (39,897 kph).

In 2004, NASA tested a hypersonic aircraft, which used a rocket booster to launch and ultimately reached speeds of 7,000 mph (10,461 kph). If this technology could successfully be applied to manned passenger flights, it would utterly change the way we move around our planet and how we experience time and distance.

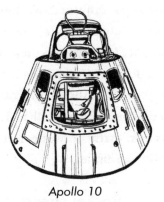

Apollo 10

Future travel

✳ Flying cars: We've all seen the sci-fi movies in which sleek, aerodynamic automobiles glide through impossibly high cityscapes. But the flying cars we may drive in the future will more likely draw on microlight technology and look a bit like two-person, enclosed gliders with detachable wings. Such vehicles will allow us to fly to nearby countries with ease. With fuel-efficient engines and being able to fly at around 150 mph without having to navigate roads, flying cars will be an attractive, environmentally friendly option. And they won't cost the Earth either – perhaps the same as a new high-end family car.

167

✳ Rubbish-powered autos: The film *Back to the Future* was made in 1985 and its closing scene featured the futuristically dressed character 'Doc' feeding waste matter into a fuel converter on this time machine/car. Back in 1985 this seemed like a pretty far-fetched idea, but today it is already a reality. Waste-to-energy plants are mushrooming across Europe, turning our unrecyclable rubbish into electricity. This electricity could soon be used to power road vehicles. While there are only a few electric cars currently on our roads, one day they'll be the norm, if hydrogen-powered cars don't beat them to it. These cars of the future will not only be greener, but considerably safer too. Traffic may even be controlled by satellite technology with vehicles talking to each other, and traffic jams could be a thing of the past.

✳ Magnetic trains: Elevated train tracks and monorails are nothing new. But magnetically levitated trains, which travel at average speeds of 260 mph, are. The first 'Maglev' train line is already in operation between the city centre in Shanghai and Pudong airport. The downside is that this new style of train transport is expensive to put in place, requiring new track. However, other train technology is catching up – and can travel nearly as fast as the Maglevs on standard tracks. Soon it'll be possible to zip from city to city in new record speeds and, as it cuts out the faff of checking in, may be even be faster than air travel in some cases.

* Slow travel: While the emphasis so far has been on speeding up, environmentalists are urging us to slow down. Largely a reaction to low-cost air travel, many in the green movement look on travelling, not just as a process of being transported from one place to another, but of experiencing the journey – preferably by taking the greener options of a train or boat. We are asked to pause and question whether we really travel any more, or do we just arrive . . . This is an entirely different approach to time – valuing the journey as much as the destination – experiencing it in real time, if you will. But as we know, real time isn't really real time – but already faster-than-human time. And it doesn't look likely to slow down any time soon.

Time travel

Humans have been fascinated by time travel for millennia. The first known story of time travel (and indeed space or inter-dimensional travel) goes back as far as the eighth century BCE in Hindu mythology. A story in the Sanskrit epic Mahabharata sees King Revaita (or Raivata) travel to a different world to meet Brahma, the god of creation, and find that many ages have passed when he returns to Earth.

Einstein's work on relativity has exerted the profoundest influence over modern thinking about time and time travel. He said that time beats at different rates depending on how fast you move. If you go fast, time slows down and this has been proven true. One experiment synchronizes two clocks, then places one in an airplane that takes off, travels around at high speed, decelerates and lands. It will be a little behind the clock that stayed on land, as the clock on the plane will operate more slowly when travelling. The difference will be small, but it will be there nonetheless. According to Einstein, both times are equally true.

Cosmonaut Sergei Krikalev holds the current record for the longest time spent in space – totalling 803 days (2.2 years) over three expeditions, the longest of which lasted 438 days. Because of the incredible speeds he was travelling (around 17,000 mph), Krikalev actually travelled into the future. In fact, he also holds the record for time travel into the future – a whopping 20 milliseconds.

These days physicists talk with considerably more confidence about the possibility of time travel – though the conditions are rather difficult to create and capture. Options include travelling at the speed of light, using cosmic strings or black holes, or prizing open tiny fissures in the space-time continuum (wormholes) and jumping into them.

Back to the beginning: finding the 'God Particle'

What is going on in the massive collider in Switzerland is way over the head of most of us mere mortals. The variable names for things and language used to describe the experiments doesn't help much either – with the terms Higgs boson, the Large Hadron Collider, CERN and the God Particle all used seemingly interchangeably. To clarify: the Large Hadron Collider (aka supercollider) at CERN (the European Laboratory for Particle Physics in

Geneva) is looking to confirm the existence of the Higgs boson (aka the God Particle).

Physicists hypothesize that the complex universe we currently know and the laws of physics that govern it evolved as the universe cooled in the first moments after the super-hot Big Bang. Now, by crashing together subatomic particles at mind-bending speed in the 17-mile circuit of the Large Hadron Collider, physicists hope to recreate and revisit the super-hot conditions of these pre-universe times to see what might have gone on back then. It is ultimately a search for original simplicity, before everything got so terribly complicated.

On the wishlist of things to find are particles, which could constitute clouds of dark matter

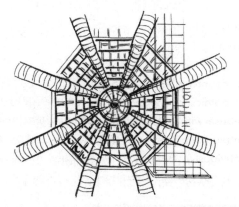

(the stuff believed to produce gravity and hold everything together and in place) and, of course, the Higgs boson – a particle which creates a sort of cosmic molasses and imbues other particles with mass. To do this, physicists are recreating the conditions of less than a billionth of a second after the universe was created, and they're doing this up to 600 million times a second. In July 2012, a new particle was found that is 'consistent with' Higgs boson but at the time of writing physicists are still reticent about giving it a firm thumbs up.

Living in the past, literally

David McDermott is an American artist who refuses to acknowledge or live in the 'present day'. He lives in mod-con-free nineteenth-century house in Dublin, Ireland, surrounded by articles of a bygone era and dressed like a country gent crossed with something out of a gothic novel, top hat and all. He refitted his house with older fixtures, fittings and furniture (though he does have a phone – an old Bakelite obviously). He says, 'I've seen the future, and I'm not going.' Refusing to use the Internet or credit, David has to physically withdraw money from the bank when he needs it.

His long-standing collaborative relationship with fellow artist Peter McGough has produced a body of painting, photography, sculpture and film that uses historical rather than modern processes in its production – as well as mixing historical eras to 'destroy the linear time system'. My personal favourite is a large painting showing a Victorian garden party against a primeval backdrop of dinosaurs and smouldering volcanoes.

Freezing time

The owner of the biggest railroad company in the US and racehorse enthusiast Leland Stanford was curious about how horses trot and wanted to know if all four feet leave the ground during the action. To find out, he recruited the photographer Eadweard Muybridge to try to capture the horse's motion in a brand-new way. In 1878, Muybridge set up twenty-four trip wires across a racetrack – capturing the motions of a galloping horse. His photographs did indeed show the horse with all four feet off the ground. He took many such series of pictures, freezing time to demonstrate the minutiae of movement. And when shown in rapid succession through a projector, we had our first 'movie'.

Leaping forward to today and experiments in freezing time have become rather more sophisticated. Harvard physicist Lene Vestergaard Hau has been conducting experiments that may pave the way for a new form of time travel – a non-human kind anyway.

Hau heats room-temperature sodium so that its atoms vibrate faster and faster. At around 350°C the atoms form a vapour. Then she forces the atoms through a pinhole and hits them with a laser beam which slows them down. This traps the atoms in an 'optical molasses' – slowing them down until they are ultimately frozen using an electromagnet. At this point, 5 to 10 million atoms are suspended in a tiny cloud, colder than any known temperature and creating a totally new state of matter. Vestergaard Hau then shoots a laser beam of light into this cold atom cloud. Consequently the light is slowed down from 186,000 miles per second to just 15 miles per hour. Once the light passes through the atom cloud it speeds up again.

Light can be slowed further and can even be stopped as if frozen in a block of ice. Not only that but Dr Hau can stop the light in one part of space and revive it in a totally different location. All the information about the light is imprinted in the atoms, creating a physical matter copy. This light can be stored indefinitely for later reactivation, thus the moment of the light is frozen in time.

Time travellers

After all that science it's time for a little science fiction. Below are my top-ten time travellers from popular culture.

10. BILL AND TED

These two guitar-playing, Californian flakes are visited by a man from a future utopia (where they are worshipped as gods) in order to help them pass a critical history test by travelling to different eras.

9. BUCK ROGERS

He started off in the 1920s and first made it onto TV in the 1950s, but the late '70s Buck Rogers is one of the most memorable time travellers around – for his ultra-tight catsuits if nothing else. An air force pilot in the year 1987, unconscious and set adrift in space for 504 years, Buck Rogers wakes up in the twenty-fifth century where he finds himself helping to defend Earth from the evil planet Draconia, with the assistance of comedy robot Twiki, and computer brain Dr Theopolis. Vintage.

8. SUPERMAN

In the first of the *Superman* films, starring Christopher Reeve (1978), we see Superman turn back time (literally) to save the woman he loves. Apparently this can be done by flying around the Earth backwards at high speed until you reverse its rotation and, therefore, time. FYI.

7. EBENEZER SCROOGE

In at seven we have everyone's favourite miser from everyone's favourite seasonal tale, *A Christmas Carol* by Charles Dickens. Scrooge leaps between past, present and future with the help of some Christmas 'ghosts' to learn lessons about generosity and love.

6. SAM BECKETT

Physicist builds time machine. Physicist tests out time machine and experiment goes pear-shaped. Physicist ends up inhabiting the bodies of men and women who lived during his lifetime to help 'put right what once went wrong' – before leaping into the next body. Physicist has holographic friend who provides him with historical data about the situations he's in. I give you Sam Beckett, lead of the gloriously implausible *Quantum Leap* (1989–93).

5. GEORGE TAYLOR

Played by Charlton Heston in the first of the *Planet of the Apes* (1968) films, George Taylor is an astronaut who takes a wrong turn and ends up propelled into the future – to find that Earth is now inhabited by intelligent, if rather bloody-minded, talking apes.

4. MARTY MCFLY

The hero of a generation, skate-boarding teenager Marty McFly of *Back to the Future* (1985) is propelled back to 1955 when trying to escape from Libyan terrorists in a time-travelling car (as you do). There he accidentally ruins the moment when his parents first meet, jeopardizing his very existence – and has to get them together in a race against the clock to get back to his own reality in 1985. In the second and third films of the franchise, Marty finds himself coming face-to-face with his future self, in a dystopic present of his own creation, and back in the Wild West.

3. THE TERMINATOR

When Arnold Schwarzenegger said 'I'll be back', he really meant it – starring in the first three films of the franchise about a cyborg killer sent from the future to the past/present to variously do away with or protect the mother of future rebel leader John Connor, then John Connor himself. Crucially, all time travel is conducted nude.

2. DOCTOR WHO

Not just a time traveller, but a Time Lord, the Doctor has been travelling through time and space in his blue police box getting into all manner of adventures since 1963. One of the longest-running TV franchises, *Doctor Who* has enchanted generations and is considered to be the most successful sci-fi series of all time.

1. THE TIME TRAVELLER

This gentleman inventor from Richmond in Surrey, England, is the central character of H. G. Wells' groundbreaking science-fiction novella *The Time Machine* (1895). The book is considered to have popularized the concept of time travel and the very term 'time machine' was coined in its pages. Testing out his new invention, the Time Traveller journeys to 802,701 CE where he encounters a race of people called the Eloi, whose conquering of technology has made them lazy, undisciplined and ultimately apathetic. Perhaps with time H. G. Wells will be seen as something of a prophet, the comparisons between the futuristic Eloi and our modern couch-potato culture are uncanny.

Time-Travel Tip
Open up a wormhole using negative energy . . . and see where it takes you

OK – so this is a little more demanding than my other tips. And chances are, if you're reading this book you're not an experimental physicist. So my advice to you is seek one out and make them your new best friend. Preferably one researching the possibility of opening wormholes using negative energy.

This area is very much in the theoretical phase – a wormhole has never been found – but who knows what might happen in the next few years? Physicists including Stephen Hawking certainly believe that they are real.

As you'll read in the next chapter, wormholes are thought to be 'shortcuts' through space and time, potentially transporting us, well, who knows where? That's if they don't close and crush us first, of course. To avoid being crushed you'd need a really fast vehicle. So far the fastest manned vehicle in history was *Apollo 10* at 25,000 mph. To travel in time through a wormhole, you'll need one that goes 2,000 times faster than that. Easy-peasy.

181

7
Space–time

Warping time

In the last chapter we touched on wormholes, tiny invisible tears in time and space, which could be portals or shortcuts to other ages and places. At their most basic, wormholes are thought to be bridges between two points in space–time. According to Stephen Hawking they are plentiful, but so small that we cannot detect them. Yet.

Wormholes crop up a lot in science fiction as they potentially allow interstellar travel within human time scales. Entering a wormhole can shave millennia off your journey time. The much-revered astrophysicist, astronomer and author Carl

Sagan (1934–1996) used wormholes as a travelling device in his novel *Contact* – in which a crew of humans make a journey to the centre of the Milky Way. Arthur C. Clarke and Stephen Baxter used wormholes for faster-than-light communication in their co-authored 2000 novel *The Light of Other Days*. And, of course, the latter-day crews of the *Star Trek* franchise frequently plunge into wormholes.

Warp speed, Mr Sulu

Wormholes aside, hyperdrives, warp drives and other such cunning inventions are the preferred methods of faster-than-light travel in science fiction. Though in a fantastical instance of life imitating art, some physicists are now saying that the warp drive idea may not be that far-fetched.

In 1994, Mexican physicist Miguel Alcubierre suggested that a real-life warp drive might be possible, though subsequent calculations have found that such a device would require prohibitive amounts of energy. Now physicists say that adjustments can be made to the proposed warp drive that would allow it to run on significantly less energy. And NASA is reportedly starting to look at the idea seriously and is conducting experiments with a mini warp drive in their laboratory at the Johnson Space Center, where they are trying to 'perturb space–time by one part in 10 million', according to Harold White who is leading the research.

A large-scale, functioning warp drive would involve a football-shaped spacecraft with a large ring encircling it. This ring would cause space–time to warp around the craft, creating a region of contracted space in front of it and expanded space behind. Meanwhile, the craft would stay inside a 'bubble' of space–time that wasn't being warped.

If humans are ever to travel truly great distances, we need to pursue such outlandish ideas in order to beat the speed of light (see below). The biggest stumbling block to exploring the wider universe is time – and the short amount of it we have to live.

Light and dark

Light travels at around 186,000 miles per second, or 671 million mph. If you cast your mind back to the section about instant messaging in the previous chapter, you'll recall that the first telegraphs travelled at around 11 million mph, taking about a quarter of a second. Well, light travels sixty-one times faster than that. The light from the Moon takes 1.3 seconds to reach us, and 8 seconds from the Sun, and four years from our next nearest star Proxima Centauri – so the light we see from it is already four years old. While we can travel faster than the speed of sound, travelling at the speed of light, let alone faster than it, is a distant dream. Unless NASA invents that warp drive, that is.

As previously mentioned, time slows down when it travels – so if we could get close to travelling at the speed of light we would age at a slower rate than the journey takes in time measured outside of the travelling craft. It is estimated that someone travelling at 99 per cent of the speed of light will age just one year in a seven-year-long journey.

Black holes

Black holes are places where gravity is so extreme that it overwhelms all other forces. Once inside, nothing can escape a black hole's gravity, not even light. Black holes are not theoretical entities. We know they exist.

Black holes are created when an object, such as a star, becomes unable to withstand the compressing force of its gravity – and the bigger the object the more gravity it has. When massive stars collapse, it is expected that they become black holes. Our Earth and Sun are too small to become black holes, but the wider universe is littered with billions of them. There is a 'supermassive' black hole at the core of our Milky Way galaxy.

Black holes do not emit any detectable light. However, astronomers can still find them. They do this by measuring visible light, X-rays and radio waves that are emitted by materials in proximity to a black hole. One way of identifying the location of black holes is by observing gas in space. If it is orbiting a black hole it tends to get very hot because of friction and then starts to emit X-rays and radio waves making the gas exceptionally bright, which can be seen using X-ray or radio telescopes. We can also detect material falling into black holes or being attracted by them.

A white hole is a hypothetical region of space–time, which is the opposite of a black hole. A white hole cannot be entered from the outside, and instead of pulling matter inside it pushes it out – matter and light.

Light years away . . .

We've all heard the expression and we know that a light year involves a very great distance indeed. But just how long is it? And how much time (as we know it) does it take to travel one?

Well, the 'simple' definition is that a light year is the distance that light travels in a vacuum during one Julian year (that's 365.25 days of 86,400 seconds each). That distance is calculated at 10 trillion kilometres (or 6 trillion miles).

This is obviously rather hard to get into perspective. If our Earth's circumference at the equator is just 24,901 miles (40,075 km), then a light year is equivalent to circumnavigating the equator 249.5 million times in a year. Or think about the distance between the Earth and Mars. At their most recent closest point in 2003, the planets were 56 million km from each other (the closest they've been in 50,000 years). That distance is only a small fraction of a light year, but in August 2012 when the Mars Science Laboratory named *Curiosity* was launched, the distance was around 100 million km and it took seven and a half months to reach the planet, yet it only takes light from Mars a few seconds to reach Earth. Is this helping with perspective?

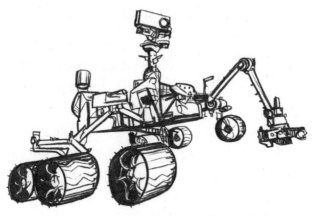

The Mars Science Laboratory, aka Curiosity

We measure in light years because we need a massive unit to make vast interstellar distances comprehensible. Our Milky Way galaxy is approximately 100,000 to 120,000 light years in diameter and contains between 200 and 400 billion stars.

The more we learn about the wider universe the larger the units of measurement we require. A trip to Mars seems like a walk in the park compared to travelling the breadth of a light year. Never mind the distances involved in parsecs (3.26 light years) or kilolight years (307 parsecs) or megalight years (307 kiloparsecs) or gigalight years (about 307 megaparsecs).

The ever-expanding universe

Now we're into the big numbers it's time to cast your mind back to the opening pages of this book. There we cast a cursory glance at the origins of our universe – the so-called Big Bang, when the universe started expanding from a tiny, dense and hot state. This event has been dated to between 13.5 and 13.75 billion years ago and the universe's expansion continues apace.

Recent data from NASA's Spitzer Space Telescope's observations of distant supernovae shows that the universe is expanding at a rate of 74.3 km per second per megaparsec (around 3 million light years) and it's speeding up.

We don't know why it's speeding up but whatever is causing it is currently being called 'dark energy'. Things we don't understand tend to be referred to as 'dark'. Dark matter is what scientists think makes up the bulk of the universe – but it can neither be seen nor detected directly with our current technologies. Over 80 per cent of our universe is believed to be comprised of this mystery material.

It is hypothesized that the continued expansion of the universe will lead to a 'big rip' – when the matter of the universe will be torn apart. Life on Earth, however, will be long over by then. It is thought that we have around 5 billion years before the Sun swallows our Earth up and burns it to a crisp.

The multiverse

The term 'multiverse' was coined by the American philosopher and psychologist William James in 1895 and refers to the hypothetical set of multiple possible universes (parallel universes/dimensions). Within the multiverse is everything that exists and everything that can exist. As appealing (or appalling) as this notion might sound, there is absolutely no proof for it and no way of testing it either.

Writing in the *New York Times* in 2003, cosmologist Paul Davies slings the worst kind of mud at the hypothesis, comparing it to religion: '. . . all cosmologists accept that there are some regions of the universe that lie beyond the reach of our telescopes, but somewhere on the slippery slope between that and the idea that there are an infinite number of universes, credibility reaches a limit. As one slips down that slope, more and more must be accepted on faith, and less and less is open to scientific verification. Extreme multiverse explanations are therefore reminiscent of theological discussions. Indeed, invoking an infinite number of unseen universes to explain the unusual features of the one we do see is just as ad hoc as invoking an unseen Creator. The multiverse theory may be dressed up in scientific language, but in essence it requires the same leap of faith.'

Parallel dimensions in popular culture

Fiction writers have been very willing to take the leap of faith required to incorporate parallel dimensions into their stories – and all the lovely paradoxes they open up. Indeed, the idea of another world parallel to our own is found in ancient tales, too – heaven and hell and their variations are parallel places. And mythic creatures tend not to roam our material world but to have access to another underworld.

Famous literary examples include *The Chronicles of Narnia* series (1950–6) by C. S. Lewis and *His Dark Materials* (1995–2000) by Philip Pullman, in which two children wander through multiple worlds, opening and closing windows between them.

Probably the most famous alternate universe portrayed on screen is Oz in the 1939 film *The Wizard of Oz*. But my personal favourite alternate-reality tale is not a science-fiction fable but rather the homely Christmas story *It's a Wonderful Life* (1946), in which the protagonist George Bailey gets to visit the hometown he has come to bitterly resent as it would have been had he never been born. He finds it a bleak and dangerous place, full of people whose lives have been stunted by his absence.

Time-Travel Tip
Get suspended

Suspended animation is still the stuff of science fiction – slowing the body's system right down into a deep stasis from which it can be reawakened unaffected by the ageing process. The best option available right now is cryonic freezing – that is, preserving your deceased, rather than live, body on ice in the hope that with advances in science it'll be possible for you to be raised from the dead to live in a hyper-advanced future.

You'll be in rather interesting company on resurrection day. There'll be James Bedford (1893–1967), a psychology professor at the University of California, who was the first man to be cryonically preserved by the Life Extension Society. There'll also be the mathematician Thomas K. Donaldson (1944–2006), computer-game designer Gregory Yob (1945–2005) and FM-2030 (1930–2000), an Iranian 'transhumanist' philosopher and writer, who'll definitely have the right kind of name in the future. Contrary to popular myth, Walt Disney will not be there: he was in fact cremated, not frozen.

8
Thinking Time

The time of our lives

Our bodies have their own internal clocks that keep them running within their own time frame – and we all experience time differently, at different times.

Subjective time

'When a man sits with a pretty woman for an hour it seems like a minute. But let him sit on a hot stove for a minute and it's longer than any hour. That's relativity.' – Albert Einstein (1879–1955).

As the Einstein quote suggests, time *feels* different depending on how we're spending it. A basic rule of thumb is that enjoyable experiences seem to

195

pass quickly, unpleasant ones more slowly. In that way our experience of time is subjective – and determined by our life experiences and expectations of past, present and future (though expert on all things spiritual, Eckhard Tolle, says nothing ever happened in the past, nor will it happen in the future – everything is now).

For some, tasks performed in their first few days at work will feel laborious and long, and then feel shorter when they have grown accustomed to them. For other people the reverse is true – even if the speed at which the tasks have been performed has been consistent.

The older one gets, one can feel that time passes more quickly – this is because many of our experiences are familiar and repeated. But cast your mind back to the long summers of your childhood – when six weeks could feel like an eternity as everything was new and exciting. In extreme or dangerous situations it can feel like time slows down, indeed, that it plays out in slow motion. And for people in prison, days drag and merge into each other because there is so little to differentiate between them.

The pace or speed at which we live depends of a great number of factors – where we live: village, town, city, country; what job we do; what hobbies we have; who our friends are, etc., etc., etc. The pace at which a trader on the New York Stock Exchange lives is rather different to that of a smallholder farmer in Kansas – though they may well keep similar hours (as we've seen in Chapter 6, money moves pretty fast these days).

The pace of life in the northern hemisphere is generally considered to be much faster than in the global South – with Switzerland and Germany singled out as the countries where the pace of life is fastest.

Biological clocks and circadian rhythms

A healthy man's heart beats around 60 times per minute throughout his adult life, and a woman's just a little faster. We breathe at pretty consistent rates too, slowing down the older we get: a newborn baby can take up to 60 breaths per minute, but an adult at rest will likely take no more than 14 to 18.

Our digestive processes and energy requirements ensure we feel hungry and need to eat at regular intervals. These are the most obviously regulated processes in our average day – but many other processes – intricate and time-dependent – are going on, collectively known as 'circadian rhythms'.

Most plants and animals in the world live to their own circadian rhythms, which largely follow a pattern dictated by hours light and dark – connecting us with our Sun. The circadian rhythm in humans is controlled by the tiny 'suprachiasmatic nucleus' in our brains. Situated on the brain's midline, behind the bridge of your nose, this nucleus is the master clock of our body. There are other 'peripheral oscillators' in our bodies too, which operate independently of the master clock and are found in the lungs, liver, pancreas and skin, and other systems.

Humans are monophasic sleepers – that is, we sleep by night and wake by day. Polyphasic sleepers indulge in multiple rest-activity cycles during a 24-hour period. It is believed that our earlier forebears were polyphasic sleepers, becoming monophasic around 70,000 to 40,000 BCE. Our circadian clocks follow this monophasic pattern.

Humans keeping regular hours will be most alert around 10 a.m. At around 2.30 p.m. our coordination is at its optimum, at 3 p.m. our reaction time is fastest. At 5 p.m. we experience our best cardiovascular efficiency and muscle strength.

At 6.30 p.m. our blood pressure is highest and at 7 p.m. our body temperature at its highest. By 9 p.m. we begin to secrete melatonin (which causes drowsiness and lowers the body temperature). Melatonin secretion decreases with age, which is why adults require less sleep than children. At 10.30 p.m. our bowel movements are suppressed. We are in our deepest slumber at 2 a.m. and lowest body temperature at 4.30 a.m., and melatonin

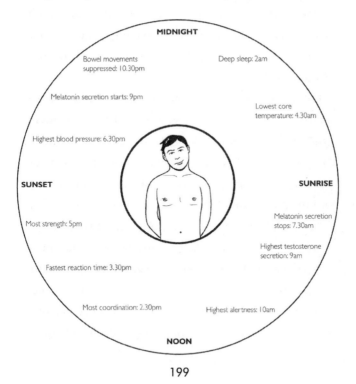

MIDNIGHT

Bowel movements suppressed: 10.30pm

Deep sleep: 2am

Melatonin secretion starts: 9pm

Lowest core temperature: 4.30am

Highest blood pressure: 6.30pm

SUNSET

SUNRISE

Most strength: 5pm

Melatonin secretion stops: 7.30am

Highest testosterone secretion: 9am

Fastest reaction time: 3.30pm

Most coordination: 2.30pm

Highest alertness: 10am

NOON

secretion stops around 7.30 a.m. as wakefulness approaches. Bowel movements commence from 8.30 a.m., with testosterone secretion reaching its highest levels at 9 a.m., and highest alertness returning again at approximately 10 a.m.

These rhythms can vary depending on the hours we keep and exposure to the Sun depending on where we live, or indeed the time of year. If we travel into different time zones and upset our internal circadian clocks, we can experience jet lag – and we need to sleep to compensate for the time difference experienced by our bodies. Sleeping on a long journey is one way of fooling your body that it has had its expected night's sleep when it arrives in a new time zone. Many people who work through the night and sleep in the day can never fully adjust to this pattern – especially as melatonin tends to be secreted at night regardless of the hours we sleep and wake.

Women have an additional 'clock' in the menstrual cycle – which comes once every twenty-eight days between a woman's teenage years and the midlife menopause.

The 28-hour day

In the 1930s, sleep researcher Nathaniel Kleitman conducted an elaborate experiment with his colleague Bruce Richardson. For thirty-two days the two lived in the Mammoth Cave in Kentucky, disrupting their circadian clocks by depriving themselves of sunlight. In addition to that, they adjusted the length of their day – living as though it was 28 rather than 24 hours long – creating a new week of six rather than seven days. They lived in a regimented way – eating, exercising and sleeping at regular times. They slept for 9 hours and were awake for 19. At forty-three years of age, Kleitman struggled to adapt to the new 28-hour, six-day week, but the younger Richardson fared better – but their results were ultimately inconclusive. Kleitman is also credited with 'discovering' Rapid Eye Movement (REM) linking dreaming and brain activity.

Accelerated ageing

Progeroid syndromes (PS) are rare genetic disorders that produce symptoms that mimic ageing. People suffering these syndromes can appear older than their actual, chronological age – and are likely to have a reduced lifespan. Werner syndrome and Hutchinson-Gilford progeria syndrome are the two most widely studied of these disorders as their effects most resemble natural ageing. The global incidence rate of Werner syndrome is one in 100,000.

People with the disorder grow normally until puberty but do not experience the expected adolescent growth spurt. Instead they exhibit a combination of growth retardation and premature ageing. They remain short, their hair grays prematurely or falls out, and their skin wrinkles. They can also experience skin atrophy, lesions, cataracts and severe ulcerations among other extreme and difficult symptoms. People afflicted with this disorder seldom live past 50 and die chiefly of cardiovascular disease or cancers.

Mating, migrating and hibernating

The behaviour of many mammals, fish and birds is inextricably linked to the seasons. Mating, migration and patterns of hibernation are all governed by the time of year and changes in the weather. Many species of birds fly south for the winter and fish such as the Atlantic salmon traverse vast distances from river to sea and back again to spawn in the streams they were born in.

To hibernate, bodies slow right down – body temperatures plummet, breathing slows to bare necessity, and heart and metabolic rates go to minimum-required function. In cold temperatures, when food is scarce, hibernators conserve energy until they can feed again. Hibernation can last days, weeks or months depending on the species. Rodents such as ground squirrels, marmots and dormice, the European hedgehog and some marsupials and primates are 'obligate' hibernators – that is, they enter hibernation annually regardless of the temperature or access to food.

Bears are among the most efficient hibernators. They rely on metabolic suppression rather than decreased body temperature to save energy during the coldest winter months and are able to recycle

their proteins and urine. They can go without 'going' for months.

Humans are one of the few creatures on the planet whose mating patterns are not dictated by the seasons. Similarly we do not hibernate, but back in our hunter-gatherer days we certainly migrated, following the patterns of our prey at different seasons. And there are still some nomadic tribal people on the planet, living their lives along ancient routes to an annual cycle.

Literary clocks

The clock has been used as a sinister plot device and metaphor in many classic tales.

'It was when I stood before her . . . that I took note of the surrounding objects in detail, and saw that her watch had stopped at 20 minutes to 9, and that a clock in the room had stopped at 20 minutes to 9.' In Charles Dickens' *Great Expectations*, the stopped clock represented the life of the eery Miss Havisham, frozen at the point that she learned that her fiancé had betrayed her on the morning of their wedding. The great clock outside the house was stopped at the same time too – as the lady herself sat in darkness, still in her wedding dress, and determined to punish the male sex for the wrongs done to her.

In her mystery novel, *The Clocks*, author Agatha Christie uses timekeepers as an elaborate plot device. When typist Sheila Webb arrives for an appointment at a house belonging to a blind lady, she finds a man lying dead in a room containing six clocks, four of which have been stopped at 4:13. Investigator extraordinaire Hercule Poirot must solve the mystery of the clocks to identify the man's killer.

In James Thurber's fantasy novel, *The Thirteen Clocks*, the eponymous timekeepers of the creepily named Coffin Castle have all been stopped at ten to five and as a consequence the megalomaniac Duke of the castle is convinced that he has conquered time. But when Prince Zorn arrives to win the hand of the Duke's niece Saralinda, their love and her great beauty make the clocks tick back to life and chime the hour of five. The couple flee and the wicked Duke gets his comeuppance. A cautionary tale for anyone who believes they can beat time's passage.

How long have we got?

For humans, how long you live depends very much on where you were born and your socio-economic circumstances. In countries generally defined as 'Western', Japan has the longest life expectancy (around 82 to 83), followed closely by Switzerland and Hong Kong (around 81 to 82). Canada, Australia, Israel and various affluent European countries including the UK, are clustered close together around the 80 bracket, while the US, where the most money is spent on individual healthcare, scores a relatively low 77.97. It is worth noting that these figures are from the UN, and considerably more generous than the World Health Organization, which scores US life expectancy from birth at 75.9 for example, while the *CIA World Factbook* gives it a generous 78.37.

According to the UN, the global average life expectancy at birth is 67.2 years (65.71 years for males and 70.14 years for females), which is pretty good considering it hovered around the 30 mark from the Bronze Age to the early twentieth century for the man in the street – and a huge number of children never made it past infancy.

Consider again this global average of 67.2 years against the poorest-performing Western countries at 76 to 78 years. There's a pretty grim reason for that ten-year disparity – and most of the countries with the lowest life expectancy are on the continent of Africa (between early 40s and late 50s), with the notable exception of Afghanistan which currently has an average life expectancy in the mid-40s.

We have a lot more time in the West. So when you're feeling stressed and musing on how little time there is to get the things you want to do done, just think of those fifty years you have on the unfortunate man from Swaziland, whom the CIA suggests might not see his thirty-second birthday.

Expressing time

In the course of writing this book I've become acutely conscious of how many expressions and phrases relate to time, as well as just how often we use the word 'time' in simple, everyday speech.

Time heals all wounds, it flies when you're having fun, it runs out on us, there's no time like the present, unless what you're doing constitutes bad timing. When things are running smoothly they're like clockwork, though it's best not to wait till the last minute or the eleventh hour or to just keep things ticking over. Saying that, it's better late than never or at the very least in the nick of time. Prisoners 'do time' and have time on their hands. Time and again, and time after time, a stitch in time saves nine.

Time to philosophize

Time and space have long fascinated our greatest thinkers. For example, do they exist independently of our minds, and do time, space and the mind exist independently of each other? Do times other than now exist concurrently with the now?

Saint Augustine of Hippo (354–430 CE) summed up the difficulty in defining and expressing time in his *Confessions*: 'If no one asks me, I know; if I seek to explain, I do not.' To him, time could only be explained by what it is not; saying what it is was another issue entirely.

How long we, and our planet, have been in existence influences our understanding of time. This book, for example, has a defined beginning point – somewhere between 13.5 and 13.75 billion years from the birth of our universe, and 4.54 billion years since the birth of our planet. But to ancient Greek philosophers there was no beginning, only an infinite impenetrable past. We've seen that later creation beliefs influenced our sense of the Earth's age – with Abrahamic religions dating our beginning to around 6,000 years ago. This was very welcome to believers as the infinite is so hard to grasp, as is nothingness and the absence of time.

Chronology and history are important to our sense of identity, but what about the nature of how we experience time and space?

Get real

Early realist philosophers believed that time and space existed separate from the human mind. Our minds are merely processors, interacting with and making sense of these external forces. Isaac Newton (1642–1727) believed that time was absolute – that there's a cosmic clock created by God that sits outside the universe and that space is the stage upon which everything happens. In his conception we have no control over time, we just have our subjective interpretations of its passage. His chief detractor was Gottfried Leibniz (1646–1716), who in the early eighteenth century challenged Newton, arguing that his 'absolutist' position did not take into consideration God's plan – there must be a specific reason why God invented time and space, if you will.

The hugely influential Immanuel Kant (1724–1804) said that the notions of time and space allow us to comprehend and coordinate our senses – but that neither have substance in themselves. To Kant, such notions are a framework we use to structure our experiences. For our purposes time and space are 'empirically real' (that is, observable), in that we use them to measure objects and experiences.

To Albert Einstein (1879–1955) time was not absolute. He thought of it as woven into the fabric of the universe and therein created. He also thought of it as something we can influence and control, as supported by the findings of many of today's physicists.

Perceiving time

Earlier we touched on the way we experience time as subjective, largely in relation to speed and pace of life. Now it's time to consider how we perceive and process it. American philosopher and psychologist William James (1842–1910) said that to live a normal life we needed a sense of 'pastness' and that our identities are constructed largely from memory and a sense of history. But this applies to much shorter time frames – our present is constantly influenced by the immediate past and often the future – as each action and thought builds on a past one towards a future one. According to James, we live in a 'specious present', rolling time frames each lasting approximately 12 seconds, and which we experience as the flow of time.

The French philosopher René Descartes (1596–1650) spoke about time being perceived as a series of instantaneous 'nows'. Similarly the contemporary spiritual teacher Eckhard Tolle (1948–) says, 'there never was a time when your life was not now'. It sounds plausible enough, but when you stop to think about it it's nigh on impossible to capture the essence of 'now'. Is now now? Or has it just past? And we know from developments in neuroscience that we don't experience or perceive things immediately, but shortly after they have occurred. It takes time for your brain to communicate information to the relevant body part, for example. It may only be half a second – but when we're talking about 'now' it all counts.

Time superstitions

Superstitions around time often relate to clocks – and may well spring from the way in which such technology encroached upon and ultimately regulated lives over the last few hundred years.

The stopped clock is perhaps the most universally known. If a clock that has stopped suddenly starts working again or chimes, it heralds a death in the family. It is also considered bad luck to stop a clock in a room in which someone has just died. These days, with clocks present in so many devices, that's quite a task.

Dreaming of clocks is meant to be prescient of an upcoming journey, while turning the hands on a clock backwards is bad luck. This likely comes from the fact that forcing the hands of older clocks with a chiming mechanism backwards can damage their workings.

In these modern times we can now predict the date of our deaths on the Internet. It is somewhat more 'scientific' than worrying about stopped clocks and takes one's date of birth, weight and body mass index into consideration. Check yours out at www.deathclock.com. I've got till October 2050 if I don't quit smoking very soon, and July 2057 if I do. Watch this space.

Time-Travel Tip
Hang out with an Amazonian tribe

Now I'm not advocating that you contact an 'uncontacted' tribe – they're best left alone. But by journeying to the Amazon, or indeed other parts of the world where the indigenous people continue to live their lives as they have been lived for millennia, it's possible to step not so much into the past, but out of time. Or out of our time at least.

Pitch a tent with the wandering Bedouins of the Middle East, meet the Maasai of East Africa or pay a respectful visit to the Amondawa in Brazil, and you will see lives lived in a different time to your own – at a completely different pace, with different rules, and different perceptions of age, past and future.

References

Books

Bryson, Bill *A Short History of Nearly Everything* (Black Swan, 2003)

Callender, Craig & Edney, Ralph *Time: A Graphic Guide* (Icon Books, 2010)

Franks, Adam *About Time* (Oneworld, 2011)

Griffiths, Jay *A Sideways Look at Time* (Tarcher, 2004)

Hart-Davis, Adam *The Book of Time* (Mitchell Beazley, 2011)

Holford-Strevens, Leofranc *The History of Time* (Oxford University Press, 2005)

Kieran, Dan *The Idle Traveller: The Art of Slow Travel* (Automobile Association, 2012)

Wilkinson, Richard & Pickett, Kate *The Spirit Level: Why Equality is Better for Everyone* (Penguin, 2010)

Articles

Davies, Paul *A Brief History of the Multiverse* (*New York Times*, 12 April 2003)

Hawking, Stephen *How to Build a Time Machine* (*Daily Mail*, 27 April 2010)

Jeffries, Stuart *The history of sleep science* (*Guardian*, 29 January 2011)

Palmer, Jason *Amondawa tribe lacks abstract idea of time, study says* (BBC News, 20 May 2011)

Radio/Podcasts

BBC Radio 4 *In Our Time: The Age of the Universe*, broadcast March 2011

BBC Radio 4 *In Our Time: The Physics of Time*, broadcast December 2008

RadioLab.org *Speed* Season 11, Episode 4, broadcast Feb 2013

RadioLab.org *Time* Season 1, Episode 4, broadcast May 2007

RadioLab.org *Beyond time* Season 1, Episode 5, July 2007

Index